TRANSACTIONS

OF THE

AMERICAN PHILOSOPHICAL SOCIETY

HELD AT PHILADELPHIA
FOR PROMOTING USEFUL KNOWLEDGE

NEW SERIES—VOLUME 55, PART 8
1965

ELEVEN-DIGIT REGULAR SEXAGESIMALS AND THEIR RECIPROCALS

OWEN GINGERICH
Smithsonian Astrophysical Observatory and Harvard University

THE AMERICAN PHILOSOPHICAL SOCIETY
INDEPENDENCE SQUARE
PHILADELPHIA 6

NOVEMBER, 1965

Library of Congress Catalog
Card Number 65-27427

ELEVEN-DIGIT REGULAR SEXAGESIMALS AND THEIR RECIPROCALS

Owen Gingerich

In their mathematics and astronomy the Babylonians employed a number system based on 60. Vestiges of this ancient arrangement still remain in our contemporary usage of minutes and seconds. The most significant feature of the sexagesimal system is its "place-value" basis, that is,

$$20°15'43'' = [20 \cdot (60)^2 + 15 \cdot (60)^1 + 43 \cdot (60)^0]''$$

Division was, whenever possible, reduced to multiplication by means of reciprocal tables. A discussion of these tables and procedures is found, for example, in the *Mathematical Cuneiform Texts* of Neugebauer and Sachs (1945). This note introduces a modern computation of such a reciprocal table.

In the decimal system, only the numbers 2 and 5, and products of their powers, have reciprocals with a finite number of digits. That is, the "regular" numbers 2, 4, 5, 8, 10, 16, 20, 25, . . . all have terminating reciprocals, whereas 3, 6, 7, . . . yield repeating fractions. The sequence of digits in the reciprocal of the regular decimal $2^p 5^q$ is the decimal representation of the integer

$$\frac{10^{p+q}}{2^p 5^q} = 2^q 5^p , \qquad (p, q \text{ positive integers}).$$

Regularity is much more common in the sexagesimal system, where 3 as well as 2 and 5 have terminating reciprocals. Numbers of the form $2^p 3^q 5^r$ (where p, q, and r are positive integers) are called *regular* sexagesimals, and the sequence of digits in their reciprocals are readily computed by multiplication as

$$\frac{60^{p+q+r}}{2^p 3^q 5^r} = 2^{p+2q+2r} 3^{p+r} 5^{p+q}$$

In each case the decimal point (or sexagesimal point) must be properly chosen to yield the correct reciprocals.

As an aid to the study of mathematical cuneiform texts, Professor Asger Aaboe[1] of Yale University suggested that I might generate an extensive table of regular sexagesimals and their reciprocals by using a modern electronic computer. Two preliminary problems presented themselves: How many entries will be found in a given order? What is the maximum number of digits in the reciprocals for a given order?

Let us define the order n as containing all sexagesi-

[1] The author wishes to thank Professor Aaboe and Professor O. Neugebauer for their suggestions in the presentation of this investigation.

mals between 60^{n-1} and 60^n. Hence the order is numerically equal to the number of sexagesimal digits. How many different combinations of p, q and r can be found such that

$$60^{n-1} \leqslant 2^p 3^q 5^r < 60^n \ ?$$

Consider the logarithmic equivalent:

$$n - 1 \leqslant p \log_{60} 2 + q \log_{60} 3 + r \log_{60} 5 < n \cdot$$

The solution may be graphically represented by plotting p, q, and r respectively along three orthogonal axes (fig. 1). The allowed combinations of p, q, and r fall in a uniformly spaced grid lying outside the plane cutting the axes at $A' = (n-1)/\log_{60} 2$, $B' = (n-1)/\log_{60} 3$, $C' = (n-1) \log_{60} 5$ but inside the plane cutting the axes at $A = (n)/\log_{60} 2$, $B = (n)/\log_{60} 3$ and $C = (n)/\log_{60} 5$. The difference in point volume between the two pyramids of points formed by the two planes can then be shown to be approximately

$$\frac{1}{6} A \cdot B \cdot C \ (1 + \tfrac{1}{2} R)^3 - \frac{1}{6} A' \cdot B' \cdot C' \ (1 + \tfrac{1}{2} R')^3$$

$$R = \frac{1}{A} + \frac{1}{B} + \frac{1}{C} \cdot \cdot$$

This formula gives close results, as seen in the following table:

Order	Formula	Actual Count
1	26	25
2	105	104
3	240	240
4	431	432
5	678	. . .
6	982	982
7	1341	. . .
8	1756	. . .
9	2227	. . .
10	2755	. . .
11	3338	3338
12	3977	. . .

The three-dimensional grid of points can be projected onto a tri-axial two-dimensional system, with all the points of a given order bounded by the projection of the face of the large pyramid. Redundant points within the small pyramid will map onto other points contained between the faces of the two pyramids. Such a representation has been discussed by O. Neugebauer (1934).

$$n \log 60/\log 2 = 5.93\,n \cdot$$

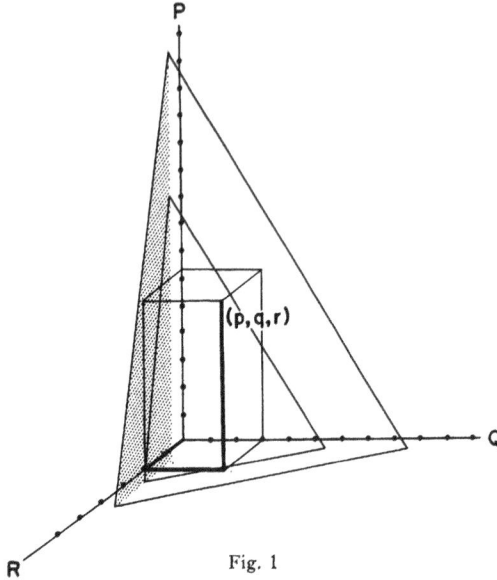

Fig. 1

The maximum number of digits in a reciprocal is readily found from the form for sexagesimal reciprocals given previously, but all factors of 60 must be removed, yielding

$$2^{2q} + {}^{2r} 3^{p/2} + r \, 5^{p/2} + q \, .$$

For the eleventh order the maximum values are a little less than $p = 65$, $q = 41$, and $r = 28$. Since $2^{82} \, 5^{41}$ is larger than either $3^{63} \, 5^{33}$ or $2^{56} \, 3^{28}$, the number having the longest reciprocal will be a little smaller than 3^{41}. The number 3^{41} has 30 sexagesimal digits, since 82 $\log_{60} 2 + 41 \log_{60} 5 = 29.9$. The actual computation of the eleventh order shows that two numbers have 30-digit reciprocals: 3^{40} and $2^1 3^{40}$. .   ‘

The internal representation of numbers in the IBM 7094 Computer used for this calculation is neither decimal nor sexagesimal; rather, binary digits or "bits" are used. Approximately six bits are required to represent each sexagesimal digit. More exactly, the number of bits required to represent an n-digit sexagesimal is

In a single computer word 35 bits are available, far too few for the production of a sexagesimal reciprocal table of useful size. In the computer's double precision mode, 54 bits are available. This would allow the production of a nine-digit sexagesimal table, which was considered too small for this project. Consequently, I adopted a scheme of paired computer words, which allowed 70 bits of numerical information. This method involved writing simple algorithms to handle the high and low order parts for addition, subtraction and multiplication. There is undoubtedly ironic justice in the fact that I had to follow the Babylonians by computing without a division operation.

Although the adopted scheme could readily handle sexagesimals of the eleventh order, the largest reciprocal would have required 174 bits for its binary representation. To avoid this, the reciprocals were obtained by successive attempted subtractions of $32 \cdot 60^n$, $16 \cdot 60^n$, $8 \cdot 60^n$, etc. In this fashion, somewhat analogous to long division, each sexagesimal digit in the reciprocal was found by a maximum of five subtractions.

At the beginning of the program, the complete set of regular sexagesimals was built up systematically, and then automatically sorted for order. Since there are 3,338 regular sexagesimals in the eleventh order, the sorting process was not formidable. The last entry in the table that follows is not counted in this total, because it duplicates the first entry. Note that the compilation is a correct reciprocal table only when the sexagesimal point is properly chosen.

Part of the computer time required for this project was purchased from the National Science Foundation grant GP 2723 to the Harvard Computing Center.

REFERENCES

NEUGEBAUER, O. 1934. *Vorlesungen über Geschichte der Antiken Mathematischen Wissenschaften. I. Vorgriechische Mathematik.* (Berlin).

———1935. *Quellen und Studien zur Geschichte der Mathematik, Astronomie und Physik,* Abteilung A: Quellen, Band 3, Mathematische Keilschift-Texte, Erster Teil, (Berlin).

NEUGEBAUER, O., AND A. SACHS. 1945. *Mathematical Cuneiform Texts,* American Oriental Series 29 (New Haven, Conn.).

P	Q	R		
0	0	0	1	60
1	16	9	1, 0, 4, 3,53,47,20,37,30	59,55,56,22,42,57,46,40
4	0	11	1, 0,16,53,53,20	59,43,1C,50,52,48
50	2	0	1, 0,19,47,28,55,18,26,33,36	59,40,19, 0,38, 5,16,15,46,29, 9,14,35,23,26,15
33	9	0	1, 0,23,52,43, 9,41,45,36	59,36,16,43,16,17,2C,37,30
16	16	0	1, 0,27,58,14, 0,57,36	59,32,14,42,18,16,17,46,40
1	24	1	1, 0,32, 4, 1,30,13,30	59,28,12,57,42,55,33,13, 5,25,44,34,53,49,37,46,40
8	0	22	1, 0,33,52,32,12,50,40,44,26,40	59,26,26,24,38,43,45,32,46,27,50,24
19	0	2	1, 0,40,53,20	59,19,34,13, 7,30
0	6	1	1, 0,45	59,15,33,20
1	22	10	1, 0,49, 6,56,42,41, 7,58, 7,3C	59,11,32,43,10,34,50,32, 5,55,33,20
23	0	13	1, 0,57,58,44,16,47,24,26,40	59, 2,56,24
0	4	10	1, 1, 2, 6,33,45	58,58,56,38,24
48	7	0	1, 1, 5, 2,19,31,59,55,23,31,12	58,56, 6,55,26,30,23,28,10,21,23,12,11,15
31	14	0	1, 1, 9,10,37,42, 4, 1,55,12	58,52, 7,37,33, 7,30
14	21	0	1, 1,13,19,12,41,28,19,12	58,48, 8,35,51,22,45,42,23,12,35,33,20
1	30	2	1, 1,17,28, 4,31,21,10, 7,30	58,44, 9,50,20,10,25,24, 2,23,56,37,25,45,18,47,34,19,15,33,20
0	2	19	1, 1,19,17,56,37, 0,18,45	58,42,24,36,11,35, 4,14,35,31,12
38	0	4	1, 1,22,14,31,54, 4,26,40	58,39,35,40,48,34,40,22,15,56,15
13	3	0	1, 1,26,24	58,35,37,30
0	12	2	1, 1,30,33,45	58,31,39,35,18,31, 6,40
11	0	8	1, 1,43,42,13,20	58,19,12
0	10	11	1, 1,47,53, 8,40,18,45	58,15,15,12
29	19	0	1, 1,55, 2,30,40,20,34,56,38,24	58, 8,31,14, 7,31,51, 6,40
12	26	0	1, 1,59,14,12, 5,59,25,26,24	58, 4,35, 9,29,15,48,50,45, 8,44, 0,19,45,11, 6,40
15	0	19	1, 2, 1, 5,19, 4, 1,58,31, 6,40	58, 2,51, 6,15,19,17,45,36
0	8	20	1, 2, 5,17,25, 4,28, 3,59, 3,45	57,58,55,24,38, 6,29,22,33,36
28	1	0	1, 2, 8,16,12,48	57,56, 8,34,22,47,34,41,15
11	8	0	1, 2,12,28,48	57,52,13,20
0	18	3	1, 2,16,41,40,18,45	57,48,18,21,32,21,50,17,17, 2,13,20
30	0	10	1, 2,25,46,13,39,45,11, 6,40	57,39,54, 8,26,15
1	1	4	1, 2,30	57,36
0	16	12	1, 2,34,14, 3,31,48,59, 3,45	57,32, 6, 7,24,26,40
10	31	0	1, 2,45,43,37,45, 3,55, 0,28,48	57,21,33,59, 0c C,48,14,34,13, 4,12,10,37,13,11,46,10,22,13,20
3	0	14	1, 2,47,36, 8, 3,20	57,19,51,12,50,41,16,48
45	0	1	1, 2,50,36,57,37,36,42,40	57,17, 6,15, 0,33,51,36,44,37,35,16,24,22,30
26	6	0	1, 2,54,52,24,57,36	57,13,13,39, 8,26,15
9	13	0	1, 2,59, 8, 9,36	57, 9,21,19, 0,44,26,40
0	24	4	1, 3, 3,24,11,33,59, 3,45	57, 5,29,14,36,24,31,53,22, 0,42,47,54, 4,26,40
18	0	5	1, 3,12,35,33,20	56,57,11,15
1	7	5	1, 3,16,52,30	56,53,2C
22	0	16	1, 3,30,23,41, 7,29,22,57,46,40	56,41,13,20,38,24
1	5	14	1, 3,34,41,50, 9,22,30	56,37,23,10,27,50,24
41	4	0	1, 3,37,44,55,20,49,55,12	56,34,4C,14,49,26,46,31,5C,44,31,52,30
24	11	0	1, 3,42, 3,34,16,19,12	56,30,5C,31,15
7	18	0	1, 3,46,22,30,43,12	56,27, 1, 3,13,19,27, 4,41,28,53,20
0	30	5	1, 3,50,41,44,42,39,33, 2,48,45	56,23,11,50,43,22, 0,23, 4,42,11, 9,31,55,30, 2,28, 8,53,20
1	3	23	1, 3,52,36,11,28,32,49,31,52,30	56,21,3C,49, 8,43,16, 4,24,29,57, 7,12
37	0	7	1, 3,55,40, 8,13,49,37,46,40	56,18,48,39,10,38, 5, 9,22,30
6	0	0	1, 4	56,15
1	13	6	1, 4, 4,20, 9,22,30	56,11,11,36,17,46,40
10	0	11	1, 4,18, 1,28,53,20	55,59,13,55,12
56	2	0	1, 4,21, 6,38,50,59,40,19,50,24	55,56,32,49,20,42,26,29,47,19,49,54,55,40,43,21,33,45
1	11	15	1, 4,22,22,51,31,59,31,52,30	55,55,26,35,31,12
39	9	0	1, 4,25,28,14, 2,20,32,38,24	55,52,45,40,34, 1,15,35, 9,22,30
22	16	0	1, 4,29,50, 6,57, 1,26,24	55,48,58,47, 9,37,46,40
5	23	0	1, 4,34,12,17,36,14,24	55,45,12, 9, 6,29,34,53,31,20,23, 2,42,57,46,40
25	0	2	1, 4,43,36,53,20	55,37, 5,49,48,16,52,30
4	5	0	1, 4,48	55,33,2C
1	19	7	1, 4,52,23,24,29,31,52,30	55,29,34,25,28,40, 9,52,35,33,20
29	0	13	1, 5, 1,50,39,13,54,34, 4,26,40	55,21,3C,22,30
0	1	7	1, 5, 6,15	55,17,45,36
37	14	0	1, 5,13,47,20,12,52,18, 2,52,48	55,11,22, 8,57,18,16,52,30
20	21	0	1, 5,18,12,29,32,14,12,28,48	55, 7,38, 3,36,55, 5,20,59,15,33,20
3	28	0	1, 5,22,37,56,49,26,34,48	55, 3,54,13,2C,24,46,18,47,14,56,50, 5,23,43,52, 5,55,33,20
2	0	17	1, 5,24,35, 8,23,28,20	55, 2,15,33,55,51,37,43,40,48
44	0	4	1, 5,27,43,30, 1,40,44,26,40	54,59,37,12, 0,32,30,20,52,26,29, 3,45
19	3	0	1, 5,32, 9,36	54,52,1C,54,22,30
2	10	0	1, 5,36,36	54,52,1C,51,51, 6,40
1	25	8	1, 5,41, 2,42, 2,54, 1,24,22,30	54,48,28, 4,25,21, 9, 0,49,55,53, 5,11, 6,40
17	0	8	1, 5,50,37, 2,13,20	54,40,3C
0	7	8	1, 5,55, 4,41,15	54,36,48
18	26	0	1, 6, 7,11, 8,54,23,23, 8, 9,36	54,26,47,57,38,41, 4,32,34,49,26,15,18,31, 6,40
1	33	0	1, 6,13,39,55,17, 3,39,44, 6	54,23, 6,53,16,27,25,44,28,53,16,52,26, 4,10,44, 2,53,23,17,31,51, 6,40
0	5	17	1, 6,13,38,34,44,45,56,15	54,21,29,26,50,43,35, 2,24
34	1	0	1, 6,16,49,17,39,12	54,18,53, 2,13,52, 6,16,10,18,45
17	8	0	1, 6,21,18,43,12	54,15,12,30
0	15	0	1, 6,25,48,27	54,11,32,12,41,35,28,23,42,13,20
36	0	10	1, 6,35,29,18,34,24,11,51, 6,40	54, 3,39,30,24,36,33,45
5	0	3	1, 6,40	54
0	13	9	1, 6,44,30,59,45,56,15	53,56,2C,44,26,40
9	0	14	1, 6,58,46,32,35,33,20	53,44,51,45,47,31,12
51	0	1	1, 7, 1,59,25,28, 7, 9,30,40	53,42,17, 6,34,44,38,11,50,14,19, 7,51, 5,37,30
0	11	18	1, 7, 3,18,48,40,49,30,42,11,15	53,41,13,31,41,57, 7,12
32	6	0	1, 7, 6,31,54,37,26,24	53,38,39, 2,56,39,36,33,45
15	13	0	1, 7,11, 4,42,14,24	53,35, 1,14, 4,26,40
0	21	1	1, 7,15,37,48,20,15	53,31,23,39,56,37,59,53,46,53,10, 7,24,26,40
24	0	5	1, 7,25,25,55,33,20	53,23,36,47,48,45
1	4	2	1, 7,30	53,20
0	19	10	1, 7,34,34,23, 0,45,42,11,15	53,16,23,26,51,31,21,28,53,20
1	2	11	1, 7,49, 0,37,30	53, 5, 2,58,33,36
47	4	0	1, 7,52,15,55, 2,13,14,52,48	53, 2,30,13,53,51,21, 7,21,19,14,52,58, 7,30
30	11	0	1, 7,56,51,48,33,24,28,48	52,58,54,51,57,46,40
13	18	0	1, 8, 1,28, 0,46, 4,48	52,55,19,44,16,14,29, 8, 8,53,20
0	27	2	1, 8, 6, 4,31,41,30,11,15	52,51,44,51,18, 9,22,51,38, 9,32,57,41,10,46,54,48,53,20
1	0	20	1, 8, 8, 6,36,14,27, 0,50	52,50,1C, 8,34,25,33,49, 7,58, 4,48
43	0	7	1, 8,11,22,48,46,44,56,17,46,40	52,47,38, 6,43,43,12,2C, 2,20,37,30
12	0	0	1, 8,16	52,44, 3,45
1	10	3	1, 8,20,37,30	52,40,29,37,46,40
16	0	11	1, 8,35,13,34,48,53,20	52,29,16,48

```
P   Q   R
1   8   12   1, 8,39,52,22,58, 7,30
45  9   0    1, 8,43,10, 6,58,29,54,48,57,36
28  16  0    1, 8,47,49,27,24,49,32, 9,36
11  23  0    1, 8,52,29, 6,46,39,21,36
0   33  3    1, 8,57, 9, 5, 5,16,18,53,26,15
1   6   21   1, 8,59,12,41,11,37,51, 5,37,30
31  0   2    1, 9, 2,31,20,53,20
10  5   0    1, 9, 7,12
1   16  4    1, 9,11,52,58, 7,30
4   0   6    1, 9,26,40
1   14  13   1, 9,31,22,17,15,21, 5,37,30
26  21  0    1, 9,39,25,19,30,23, 9,18,43,12
9   28  0    1, 9,44, 8,28,36,44,21, 7,12
8   0   17   1, 9,46,13,28,57, 2,13,20
50  0   4    1, 9,49,34,24, 1,47,27,24,26,40
25  3   0    1, 9,54,18,14,24
8   10  0    1, 9,59, 2,24
1   22  5    1,10, 3,46,52,51, 5,37,30
23  0   8    1,10,13,59,30,22,13,20
0   4   5    1,10,18,45
7   33  0    1,10,36,26,34,58,11,54,23, 2,24
0   2   14   1,10,38,33, 9, 3,45
40  1   0    1,10,41,56,34,49,48,48
23  8   0    1,10,46,43,58, 4,48
6   15  0    1,10,51,31,40,48
1   28  6    1,10,56,19,43, 0,43,56,43, 7,30
0   0   23   1,10,58,26,52,45, 3, 8,22, 5
11  0   3    1,11, 6,40
0   10  6    1,11,11,29, 3,45
15  0   14   1,11,26,41,38,45,55,33,20
57  0   1    1,11,30, 7,23, 9,59,38, 8,42,40
0   8   15   1,11,31,32, 3,55,32,48,45
38  6   0    1,11,34,58, 2,15,56, 9,36
21  13  0    1,11,39,49, 1, 3,21,36
4   20  0    1,11,44,40,19,33,36
30  0   5    1,11,55, 7,39,15,33,20
3   2   0    1,12
0   16  7    1,12, 4,52,40,32,48,45
3   0   9    1,12,20,16,40
53  4   0    1,12,23,44,58,42,22, 7,52,19,12
0   14  16   1,12,25,10,42,58,29,28,21,23,45
36  11  0    1,12,28,39,15,47,38, 6,43,12
19  18  0    1,12,33,33,52,49, 9, 7,12
2   25  0    1,12,38,28,49,48,16,12
7   0   20   1,12,40,39, 2,39,24,48,53,20
18  0   0    1,12,49, 4
1   7   0    1,12,54
0   22  8    1,12,58,56,20, 3,13,21,33,45
22  0   11   1,13, 9,34,29, 8, 8,53,20
1   5   9    1,13,14,31,52,30
34  16  0    1,13,23, 0,45,14,28,50,18,14,24
17  23  0    1,13,27,59, 3,13,45,59, 2,24
0   30  0    1,13,32,57,41,25,37,24, 9
1   3   18   1,13,35, 9,31,56,24,22,30
37  0   2    1,13,38,41,26,16,53,20
16  5   0    1,13,43,40,48
1   13  1    1,13,48,40,30
10  0   6    1,14, 4,26,40
1   11  10   1,14, 9,27,46,24,22,30
15  28  0    1,14,23, 5, 2,31,11,18,31,40,48
14  0   17   1,14,25,18,22,52,50,22,13,20
0   36  1    1,14,28, 7,24,41,41,37,12, 6,45
1   9   19   1,14,30,20,54, 5,21,40,46,52,30
31  3   0    1,14,33,55,27,21,36
14  10  0    1,14,38,58,33,36
1   19  2    1,14,44, 2, 0,22,30
29  0   8    1,14,54,55,28,23,42,13,20
0   1   2    1,15
1   17  11   1,15, 5, 4,52,14,10,46,52,30
2   0   12   1,15,21, 7,21,40
46  1   0    1,15,24,44,21, 9, 8, 3,12
29  8   0    1,15,29,50,53,57, 7,12
12  15  0    1,15,34,57,47,31,12
1   25  3    1,15,40, 5, 1,52,46,52,30
6   0   23   1,15,42,20,40,16, 3,20,55,33,20
17  0   3    1,15,51, 6,40
0   7   3    1,15,56,15
21  0   14   1,16,12,28,25,20,59,15,33,20
0   5   12   1,16,17,38,12,11,15
44  6   0    1,16,21,17,54,24,59,54,14,24
27  13  0    1,16,26,28,17, 7,35, 2,24
10  20  0    1,16,31,39, 0,51,50,24
1   31  4    1,16,36,50, 5,39,11,27,39,22,30
0   3   21   1,16,39, 7,25,46,15,23,26,15
36  0   5    1,16,42,48, 9,52,35,33,20
9   2   0    1,16,48
0   13  4    1,16,53,12,11,15
9   0   9    1,17, 9,37,46,40
0   11  13   1,17,14,51,25,50,23,26,15
42  11  0    1,17,18,33,52,50,48,39,10, 4,48
25  18  0    1,17,23,48, 8,20,25,43,40,48
8   25  0    1,17,29, 2,45, 7,29,16,48
13  0   20   1,17,31,21,38,50, 2,28, 8,53,20
24  0   0    1,17,40,20,16
7   7   0    1,17,45,36
0   19  5    1,17,50,52, 5,23,26,15
28  0   11   1,18, 2,12,47, 4,41,28,53,20
1   2   6    1,18, 7,30
0   17  14   1,18,12,47,34,24,46,13,49,41,15
23  23  0    1,18,21,50,59,26,41, 2,58,33,36
```

```
52,25,43,40,48
52,23,12,49,16,53,40,51,42,32,20,37,30
52,19,4C, 6,42,46,40
52,16, 7,38,32,20,13,57,40,37,51,36,17,46,40
52,12,35,24,44,35,55,54,42, 7,56,59,56,13,36,42,17,10,27, 9,37,46,40
52,11, 1,52,10,17,5C,26,18,14,24
52, 8,31,42,56,30,49,13, 7,30
52, 5
52, 1,28,31,23, 7,39,15,33,20
51,50,24
51,46,53,30,40
51,40,54,25,53,21,38,45,55,33,20
51,37,24,35, 6, 0,43,25, 6,47,45,46,57,13,29,52,35,33,20
51,35,52, 5,33,37, 9, 7,12
51,33,23,37,3C,30,28,27, 4, 9,49,44,45,56,15
51,29,54,17,13,35,37,30
51,26,25,11, 6,40
51,22,56,19, 8,46, 4,42, 1,48,38,31, 6,40
51,15,28, 7,30
51,12
50,59,10,12,26,40,42,52,57, 4,57, 4, 9,26,25, 3,47,42,33, 5,11, 6,40
50,57,36,51,25, 3,21,36
50,55,12,13,20,30, 5,52,39,40, 4,41,15
50,51,45,28, 7,30
50,48,18,56,53,59,30,22,13,20
50,44,52,39,39, 1,48,20,46,13,58, 2,34,43,57, 2,13,20
50,43,21,44,13,50,56,27,58, 2,57,24,28,48
50,37,3C
50,34, 4,26,40
5C,23,18,31,40,48
50,20,53,32,24,38,11,50,48,35,50,55,26, 6,39, 1,24,22,30
50,19,53,55,58, 4,48
50,17,29, 6,30,37, 8, 1,38,26,15
50,14, 4,54,26,40
50,10,40,56,11,50,37,24,1C,12,20,44,26,40
50, 3,23,14,49,27,11,15
50
49,56,3E,58,55,48, 8,53,20
49,45,59, 2,24
49,43,35,50,31,44,23,33, 8,44,17,42, 9,29,31,52,30
49,42,36,58,14,24
49,40,13,56, 3,34,27,11,15
49,36,52,15,15,13,34,48,53,20
49,33,3C,48, 5,46,17,40,54,31,27, 9, 4,51,21,28,53,20
49,32, 2, 0,32,16,27,57,18,43,12
49,26,18,30,56,15
49,22,57,46,40
49,19,37,15,58,49, 2, 6,44,56,17,46,40
49,12,27
49, 9, 7,12
49, 3,26,21,17,36,15
49, 0, 7, 9,52,48,58, 5,19,20,29,37,46,40
48,56,48,11,56,48,41,10, 1,59,57,11,11,27,45,39,38,36, 2,57,46,40
48,55,2C,30, 9,39,36
48,52,59,44, 0,28,53,38,33,16,52,30
48,49,41,15
48,46,22,59,25,25,55,33,20
48,36
48,32,42,40
48,23,49,17,54,23,10,42,17,37,16,40,16,27,39,15,33,20
48,22,22,35,12,46, 4,48
48,20,32,47,21,17,42,52,52,20,41,39,56,30,22,52,29,14, 7,22,14,58,45,55,33,20
48,19, 6,10,31,45,24,28,48
48,16,47, 8,38,59,38,54,22,30
48,13,31, 6,40
48,10,15,17,56,58,11,54,24,11,51, 6,40
48, 3,15, 7, 1,52,30
48
47,56,45, 6,10,22,13,20
47,46,32,40,42,14,24
47,44,15,12,30,28,13, 0,37,11,19,23,40,18,45
47,41, 1,22,37, 1,52,30
47,37,47,45,50,37, 2,13,20
47,34,34,22,10,20,26,34,28,20,35,39,55, 3,42,13,20
47,27,39,22,30
47,24,26,40
47,14,21, 7,12
47,11, 9,18,43,12
47, 8,53,32,21,12,18,46,32,17, 6,33,45
47, 5,42, 6, 2,30
47, 2,30,52,41, 6,12,33,54,34, 4,26,40
46,59,19,52,16, 8,2C,19,13,55, 9,17,56,36,15, 2, 3,27,24,26,40
46,57,55,40,57,16, 3,23,40,24,57,36
46,55,40,32,38,51,44,17,48,45
46,52,3C
46,49,19,40,14,48,53,20
46,39,21,36
46,36,12, 9,36
46,33,58, 3,48,21, 2,59,17,48,45
46,30,48,59,18, 1,28,53,20
46,27,4C, 7,35,24,39, 4,36, 6,59,12,15,48, 8,53,20
46,26,16,53, 0,15,26,12,28,48
46,20,54,51,30,14, 3,45
46,17,46,40
46,14,38,41,13,53,28,13,49,37,46,40
46, 7,55,18,45
46, 4,48
46, 1,4C,53,55,33,2C
45,56,21,43, 0,45,54,27,29,22,57,46,40
```

P	Q	R		
6	30	0	1,18,27, 9,32,11,19,53,45,36	45,53,15,11,12, 0,38,35,39,22,27,21,44,29,46,33,24,56,17,46,40
1	0	15	1,18,29,30,10, 4,10	45,51,52,58,16,33, 1,26,24
43	0	2	1,18,33,16,12, 2, 0,53,20	45,49,41, 0, 0,27, 5,17,23,42, 4,13, 7,30
22	5	0	1,18,38,35,31,12	45,46,34,55,18,45
5	12	0	1,18,43,55,12	45,43,26, 3,12,35,33,20
0	25	6	1,18,49,15,14,27,28,49,41,15	45,40,23,23,41, 7,37,30,41,36,34,14,19,15,33,20
16	0	6	1,19, 0,44,26,40	45,33,45
1	8	7	1,19, 6, 5,37,30	45,30,4C
20	0	17	1,19,22,59,36,24,21,43,42,13,20	45,20,58,40,30,43,12
4	35	0	1,19,25,59,54,20,28,23,40,55,12	45,19,15,44,23,42,51,27, 4, 4,24, 3,41,43,28,56,42,24,29,24,36,32,35,33,20
1	6	16	1,19,28,22,17,41,43, 7,30	45,17,54,32,22,16,19,12
37	3	0	1,19,32,11, 9,11, 2,24	45,15,44,11,51,33,25,13,28,35,37,30
20	10	0	1,19,37,34,27,50,24	45,12,4C,25
3	17	0	1,19,42,58, 8,24	45, 9,36,50,34,39,33,39,45,11, 6,40
35	0	8	1,19,54,35,10,17,17, 2,13,20	45, 3, 2,55,20,30,28, 7,30
4	0	1	1,20	45
1	14	8	1,20, 5,25,11,43, 7,30	44,56,57,17, 2,13,2C
8	0	12	1,20,22,31,51, 6,40	44,47,23, 8, 9,36
52	1	0	1,20,26,23,18,33,44,35,24,48	44,45,14,15,28,33,57,11,49,51,51,55,56,32,34,41,15
1	12	17	1,20,27,58,34,24,59,24,50,37,30	44,44,21,16,24,57,36
35	8	0	1,20,31,50,17,32,55,40,48	44,42,12,32,27,13, 0,28, 7,30
18	15	0	1,20,37,17,38,41,16,48	44,39,11, 1,43,42,13,20
1	22	0	1,20,42,45,22, 0,18	44,36, 9,43,17,11,39,54,49, 4,18,26,10,22,13,20
23	0	3	1,20,54,31, 6,40	44,29,4C,39,50,37,3C
0	4	0	1,21	44,26,4C
1	20	9	1,21, 5,29,15,36,54,50,37,30	44,23,39,32,22,56, 7,54, 4,26,40
27	0	14	1,21,17,18,19, 2,23,12,35,33,20	44,17,12,18
0	2	9	1,21,22,48,45	44,14,12,28,48
50	6	0	1,21,26,43, 6, 2,39,53,51,21,36	44,12, 5,11,34,52,47,36, 7,46, 2,24, 8,26,15
33	13	0	1,21,32,14,10,16, 5,22,33,36	44, 9, 5,43, 9,50,37,30
16	20	0	1,21,37,45,36,55,17,45,36	44, 6, 6,26,53,32, 4,16,47,24,26,40
1	28	1	1,21,43,17,26, 1,48,13,30	44, 3, 7,22,45, 7,49, 3, 1,47,57,28, 4,18,59, 5,40,44,26,40
0	0	18	1,21,45,43,55,29,20,25	44, 1,48,27, 8,41,18,10,56,38,24
42	0	5	1,21,49,39,22,32, 5,55,33,20	43,59,41,45,36,26, 0,16,41,57,11,15
15	2	0	1,21,55,12	43,56,43, 7,30
0	10	1	1,22, 0,45	43,53,44,41,28,53,20
15	0	9	1,22,18,16,17,46,40	43,44,24
0	8	10	1,22,23,50,51,33,45	43,41,26,24
31	18	0	1,22,33,23,20,53,47,26,35,31,12	43,36,23,25,35,38,53,20
14	25	0	1,22,38,58,56, 7,59,13,55,12	43,33,26,22, 6,56,51,38, 3,51,33, 0,14,48,53,20
1	34	2	1,22,44,34,54, 6,19,34,40, 7,30	43,30,29,30,37, 9,56,35,35, 6,37,29,56,51,20,35,14,18,42,38, 1,28,53,20
0	6	19	1,22,47, 3,13,25,57,25,18,45	43,29,11,33,28,34,52, 1,55,12
30	0	0	1,22,51, 1,37, 4	43,27, 6,25,47, 5,41, 0,56,15
13	7	0	1,22,56,38,24	43,24,10
0	16	2	1,23, 2,15,33,45	43,21,13,46, 9,16,22,42,57,46,40
34	0	11	1,23,14,21,38,13, 0,14,48,53,2C	43,14,55,36,19,41,15
3	0	4	1,23,20	43,12
0	14	11	1,23,25,38,44,42,25,18,45	43, 9, 4,35,33,20
12	30	0	1,23,40,58,10,20, 5,13,20,38,24	43, 1,10,29,15, 0,36,10,55,39,48, 9, 7,57,54,53,49,37,46,40
7	0	15	1,23,43,28,10,44,26,40	42,59,53,24,38, 0,57,36
49	0	2	1,23,47,29,16,50, 8,56,53,20	42,57,49,41,15,25,23,42,33,28,11,27,18,16,52,30
28	5	0	1,23,53, 9,53,16,48	42,54,55,14,21,19,41,15
11	12	0	1,23,58,50,52,48	42,52, C,59,15,33,2C
0	22	3	1,24, 4,32,15,25,18,45	42,49, 6,55,57,18,23,55, 1,30,32, 5,55,33,20
22	0	6	1,24,16,47,24,26,40	42,42,53,26,15
1	5	4	1,24,22,30	42,40
0	20	12	1,24,28,12,58,45,57, 7,44, 3,45	42,37, 6,45,29,13, 5,11, 6,40
1	3	13	1,24,46,15,46,52,30	42,28, 2,22,50,52,48
43	3	0	1,24,50,19,53,47,46,33,36	42,26, C,11, 7, 5, 4,53,53, 3,23,54,22,30
26	10	0	1,24,56, 4,45,41,45,36	42,23, 7,53,26,15
9	17	0	1,25, 1,50, 0,57,36	42,20,15,47,24,59,35,18,31, 6,40
0	28	4	1,25, 7,35,39,36,52,44, 3,45	42,17,23,53, 2,31,30,17,18,31,38,22, 8,56,37,31,51, 6,40
1	1	22	1,25,10, 8,15,18, 3,46, 2,30	42,16, 8, 6,51,32,27, 3,18,22,27,50,24
41	0	8	1,25,14,13,30,58,26,10,22,13,20	42,14, 6,29,22,58,33,52, 1,52,30
10	0	1	1,25,20	42,11,15
1	11	5	1,25,25,46,52,30	42, 8,23,42,13,20
14	0	12	1,25,44, 1,58,31, 6,40	41,59,25,26,24
58	1	0	1,25,48, 8,51,47,59,33,46,27,12	41,57,24,37, 0,31,49,52,20,29,52,26,11,45,32,31,10,18,45
1	9	14	1,25,49,50,28,42,39,22,30	41,56,34,56,38,24
41	8	0	1,25,53,57,38,43, 7,23,31,12	41,54,34,15,25,30,56,41,22, 1,52,30
24	15	0	1,25,59,46,49,16, 1,55,12	41,51,44, 5,22,13,20
7	22	0	1,26, 5,36,23,28,19,12	41,48,54, 6,49,52,11,10, 8,30,17,17, 2,13,20
29	0	3	1,26,18, 9,11, 6,40	41,42,49,22,21,12,39,22,30
6	4	0	1,26,24	41,40
1	17	6	1,26,29,51,12,39,22,30	41,37,10,49, 6,30, 7,24,26,40
2	0	7	1,26,48,20	41,28,19,12
1	15	15	1,26,54,12,51,34,11,22, 1,52,3C	41,25,3C,48,32
39	13	0	1,26,58,23, 6,57, 9,44, 3,50,24	41,23,31,36,42,58,42,39,22,30
22	20	0	1,27, 4,16,39,22,58,56,38,24	41,20,42,32,42,41,19, 0,44,26,40
5	27	0	1,27,10,10,35,45,55,26,24	41,17,55,40, 4,48,34,44, 5,26,12,37,34, 2,47,54, 4,26,40
6	0	18	1,27,12,46,51,11,17,46,40	41,16,41,40,26,53,43,17,45,36
48	0	5	1,27,16,58, 0, 2,14,19,15,33,20	41,14,42,54, 0,24,22,45,39,19,51,47,48,45
21	2	0	1,27,22,52,48	41,11,55,25,46,52,30
4	9	0	1,27,28,48	41, 9, 8, 8,53,20
1	23	7	1,27,34,43,36, 3,52, 1,52,30	41, 6,21, 3,19, 0,51,45,37,26,54,48,53,20
21	0	9	1,27,47,29,22,57,46,40	41, 0,22,30
0	5	7	1,27,53,26,15	40,57,36
20	25	0	1,28, 9,34,51,52,31,10,50,52,48	40,50, 5,58,14, 0,48,24,26, 7, 4,41,28,53,20
3	32	0	1,28,15,33,13,42,44,52,58,48	40,47,2C, 9,57,20,34,18,21,39,57,39,19,33, 8, 3, 2,10, 2,28, 8,53,20
0	3	16	1,28,18,11,26,19,41,15	40,46, 7, 5, 8, 2,41,16,48
36	0	0	1,28,22,25,43,32,16	40,44, 9,46,40,24, 4,42, 7,44, 3,45
19	7	0	1,28,28,24,57,36	40,41,24,22,30
2	14	0	1,28,34,24,36	40,38,39, 9,31,11,36,17,46,40
0	1	25	1,28,43, 3,35,56,18,55,27,36,15	40,34,41,23,23, 4,45,10,22,26,21,55,35, 2,24
9	0	4	1,28,53,20	40,30
0	11	8	1,28,59,21,19,41,15	40,27,15,33,20
13	0	15	1,29,18,22, 3,27,24,26,40	40,18,38,49,20,38,24
1	37	0	1,29,21,44,53,38, 1,56,38,32, 6	40,17, 7,19,27,44,45,44, 3,37,14,43,17, 5,19, 3,44,21,46, 8,32,28,58,16,17,46,40
55	0	2	1,29,22,39,13,57,29,32,40,53,20	40,16,42,49,55,42,33,28,38,52,40,44,20,53,19,13, 7,30
0	9	17	1,29,24,25, 4,54,26, 0,56,15	40,15,55, 8,46,27,50,24

```
 P  Q  R
34  5  0   1,29,28,42,32,49,55,12          40,13,55,17,12,29,42,25,18,45
17 12  0   1,29,34,46,16,19,12             40,11,15,55,33,20
 0 19  0   1,29,40,50,24,27                40, 8,32,44,57,28,29,55,20, 9,52,35,33,20
28  0  6   1,29,53,54,34, 4,26,40          40, 2,42,35,51,33,45
 1  2  1   1,30                            40
 0 17  9   1,30, 6, 5,50,41, 0,56,15       39,57,17,35, 8,38,31, 6,40
 1  0 10   1,30,25,20,50                   39,48,47,13,55,12
49  3  0   1,30,29,41,13,22,57,39,50,24    39,46,52,40,25,23,30,50,30,59,26, 9,43,35,37,30
32 10  0   1,30,35,49, 4,44,32,38,24       39,44,11, 8,50,51,33,45
15 17  0   1,30,41,57,21, 1,26,24          39,41,29,48,12,10,51,51, 6,40
 0 25  1   1,30,48, 6, 2,15,20,15          39,38,48,38,28,37, 2, 8,43,37, 9,43,15,53, 5,11, 6,40
 5  0 21   1,30,50,48,48,19,16, 1, 6,40    39,37,37,36,25,49,10,21,50,58,33,36
16  0  1   1,31, 1,20                      39,33, 2,48,45
 1  8  2   1,31, 7,30                      39,30,22,13,20
 0 23 10   1,31,13,40,25, 4, 1,41,57,11,15 39,27,41,48,47, 3,13,41,23,57, 2,13,20
20  0 12   1,31,26,58, 6,25,11, 6,40       39,21,57,36
 1  6 11   1,31,33, 9,50,37,30             39,19,17,45,36
47  8  0   1,31,37,33,29,17,59,53, 5,16,48 39,17,24,36,57,40,15,38,46,54,15,28, 7,30
30 15  0   1,31,43,45,56,33, 6, 2,52,48    39,14,45, 5, 2, 5
13 22  0   1,31,49,58,49, 2,12,28,48       39,12, 5,43,54,15,10,28,15,28,23,42,13,70
 0 31  2   1,31,56,12, 6,47, 1,45,11,15    39, 9,26,33,33,26,56,56, 1,35,57,44,57,10,12,31,42,52,50,22,13,20
 1  4 20   1,31,58,56,54,55,30,28, 7,30    39, 8,16,24, 7,43,22,49,43,40,48
35  0  3   1,32, 3,21,47,51, 6,40          39, 6,23,47,12,23, 6,54,50,37,30
12  4  0   1,32, 9,36                      39, 3,45
 1 14  3   1,32,15,50,37,30                39, 1, 6,23,32,20,44,26,40
 8  0  7   1,32,35,33,20                   38,52,48
 1 12 12   1,32,41,49,43, 0,28, 7,30       38,50,10, 8
28 20  0   1,32,52,33,46, 0,30,52,24,57,36 38,45,40,49,25, 1,14, 4,26,40
11 27  0   1,32,58,51,18, 8,59, 8, 9,36    38,43, 3,26,19,30,32,33,50, 5,49,20,13,10, 7,24,26,40
12  0 18   1,33, 1,37,58,36, 2,57,46,40    38,41,54, 4,10,12,51,50,24
27  2  0   1,33,12,24,19,12                38,37,25,42,55,11,43, 7,30
10  9  0   1,33,18,43,12                   38,34,48,53,20
 1 20  4   1,33,25, 2,30,28, 7,30          38,32,12,14,21,34,33,31,31,21,28,53,20
27  0  9   1,33,38,39,20,29,37,46,40       38,26,36, 5,37,30
 0  2  4   1,33,45                         38,24
 1 18 13   1,33,51,21, 5,17,43,28,35,37,30 38,21,24, 4,56,17,46,40
 9 32  0   1,34, 8,35,26,37,35,52,30,43,12 38,14,22,39,20, 0,32, 9,42,48,42,48, 7, 4,48,47,50,46,54,48,53,20
 0  0 13   1,34,11,24,12, 5                38,13,14, 8,33,47,31,12
42  0  0   1,34,15,55,26,26,25, 4          38,11,24,10, 0,22,34,24,29,45, 3,30,56,15
25  7  0   1,34,22,18,37,26,24             38, 8,49, 6, 5,37,30
 8 14  0   1,34,28,42,14,24                38, 6,14,12,40,29,37,46,40
 1 26  5   1,34,35, 6,17,20,58,35,37,30    38, 3,35,29,44,16,21,15,34,40,28,31,56, 2,57,46,40
 4  0 24   1,34,37,55,50,20, 4,11, 9,26,40 38, 2,31,18,10,23,12,20,58,32,13, 3,21,36
15  0  4   1,34,48,53,20                   37,58, 7,30
 0  8  5   1,34,55,18,45                   37,55,33,20
19  0 15   1,35,15,35,31,41,14, 4,26,40    37,47,28,53,45,36
 0  6 14   1,35,22, 2,45,14, 3,45          37,44,55,26,58,33,36
40  5  0   1,35,26,37,23, 1,14,52,48       37,43, 6,49,52,57,51, 1,13,49,41,15
23 12  0   1,35,33, 5,21,24,28,48          37,40,33,40,50
 6 19  0   1,35,39,33,46, 4,48             37,38, 0,42, 8,52,58, 3, 7,39,15,33,20
 0  4 23   1,35,48,54,17,12,49,14,17,48,45 37,34,20,32,45,48,50,42,56,19,58, 4,48
34  0  6   1,35,53,30,12,20,44,26,40       37,32,32,26, 7, 5,23,26,15
 5  1  0   1,36                            37,30
 0 14  6   1,36, 6,30,14, 3,45             37,27,27,44,11,51, 6,40
 7  0 10   1,36,27, 2,13,20                37,19,25,16,48
55  3  0   1,36,31,39,58,16,29,30,29,45,36 37,17,41,52,53,48,17,39,51,33,13,16,37, 7, 8,54,22,30
 0 12 15   1,36,33,34,17,17,59,17,48,45    37,16,57,43,40,48
38 10  0   1,36,38,12,21, 3,30,48,57,36    37,15,10,27, 2,40,50,23,26,15
21 17  0   1,36,44,45,10,25,32, 9,36       37,12,35,11,26,25,11, 6,40
 4 24  0   1,36,51,18,26,24,21,36          37,10, 8, 6, 4,19,43,15,40,53,35,21,48,38,31, 6,40
11  0 21   1,36,54,12, 3,32,33, 5,11, 6,40 37, 9, 1,30,24,12,20,57,59, 2,24
22  0  1   1,37, 5,25,20                   37, 4,43,53,12,11,15
 3  6  0   1,37,12                         37, 2,13,20
 0 20  7   1,37,18,35, 6,44,17,48,45       36,59,42,56,59, 6,46,35, 3,42,13,20
26  0 12   1,37,32,45,58,50,51,51, 6,40    36,54,20,15
 1  3  8   1,37,39,22,30                   36,51,50,24
36 15  0   1,37,50,41, 0,19,18,27, 4,19,12 36,47,34,45,58,12,11,15
19 22  0   1,37,57,18,44,18,21,18,43,12    36,45, 5,22,24,36,43,33,59,30,22,13,20
 2 29  0   1,38, 3,56,55,14, 9,52,12       36,42,36, 8,57,36,30,52,31,29,57,53,23,35,49,14,43,57, 2,13,20
 1  1 17   1,38, 6,52,42,35,12,30          36,41,30,22,37,14,25, 9, 7,12
41  0  3   1,38,11,35,15, 2,31, 6,40       36,39,44,48, 0,21,40,13,54,57,39,22,30
18  4  0   1,38,18,14,24                   36,37,15,56,15
 1 11  0   1,38,24,54                      36,34,47,14,34, 4,26,40
 0 26  8   1,38,31,34, 3, 4,21, 2, 6,33,45 36,32,18,42,56,54, 6, 0,33,17,15,23,27,24,26,40
14  0  7   1,38,45,55,33,20                36,27
 1  9  9   1,38,52,37, 1,52,30             36,24,32
17 27  0   1,39,10,46,43,21,35, 4,42,14,24 36,17,51,58,25,47,23, 1,43,12,57,30,12,20,44,26,40
18  0 18   1,39,13,44,30,30,27, 9,37,46,40 36,16,46,56,24,34,33,36
 0 34  0   1,39,17,29,52,55,35,29,36, 9    36,15,24,35,30,58,17, 9,39,15,31,14,57,22,47, 9,21,55,35,31,41,14, 4,26,40
 1  7 18   1,39,20,27,52, 7, 8,54,22,30    36,14,19,37,53,49, 3,21,36
33  2  0   1,39,25,13,56,28,48             36,12,35,21,29,14,44,10,46,52,30
16  9  0   1,39,31,58, 4,48                36,10, 8,20
 1 17  1   1,39,38,42,40,30                36, 7,41,28,27,43,38,55,48, 8,53,20
33  0  9   1,39,53,13,57,51,36,17,46,40    36, 2,26,20,16,24,22,30
 2  0  2   1,40                            36
 1 15 10   1,40, 6,46,29,38,54,22,30       35,57,33,49,37,46,40
 6  0 13   1,40,28, 9,48,53,20             35,49,54,30,31,40,48
48  0  0   1,40,32,59, 8,12,10,44,16       35,48,11,24,22,51, 9,45,27,53,29,32,45,14, 3,45
31  7  0   1,40,39,47,51,56, 9,36          35,45,40, 1,57,46,24,22,30
14 14  0   1,40,46,37, 3,21,36             35,43,20,49,22,57,46,40
 1 23  2   1,40,53,26,42,30,22,30          35,40,55,46,37,45,19,55,51,15,26,44,56,17,46,40
21  0  4   1,41, 8, 8,53,20                35,35,44,31,52,30
 0  5  2   1,41,15                         35,33,20
 1 21 11   1,41,21,51,34,31, 8,33,16,52,30 35,30,55,37,54,20,54,19,15,33,20
25  0 15   1,41,36,37,53,47,59, 0,44,26,40 35,25,45,50,24
 0  3 11   1,41,43,30,56,15                35,23,21,59, 2,24
46  5  0   1,41,48,23,52,33,19,52,19,12    35,21,40, 9,15,54,14, 4,54,12,49,55,18,45
29 12  0   1,41,55,17,42,50, 6,43,12       35,19,16,34,31,52,30
12 19  0   1,42, 2,12, 1, 9, 7,12          35,16,53, 9,30,49,39,25,25,55,33,20
 1 29  3   1,42, 9, 6,47,32,15,16,52,30    35,14,29,54,12, 6,15,14,25,26,21,58,27,27,11,16,32,35,33,20
```

P	Q	R		
0	1	20	1,42,12, 9,54,21,40,31,15	35,13,26,45,42,57, 2,32,45,18,43,12
40	0	6	1,42,17, 4,13,10, 7,24,26,40	35,11,45,24,29, 8,48,13,21,33,45
11	1	0	1,42,24	35, 9,22,30
0	11	3	1,42,30,56,15	35, 6,59,45,11, 6,40
13	0	10	1,42,52,50,22,13,20	34,59,31,12
0	9	12	1,42,59,48,34,27,11,15	34,57, 5, 7,12
44	10	0	1,43, 4,45,10,27,44,52,13,26,24	34,55,28,32,51,15,47,14,28,21,33,45
27	17	0	1,43,11,44,11, 7,14,18,14,24	34,53, 6,44,28,31, 6,40
10	24	0	1,43,18,43,40, 9,59, 2,24	34,50,45, 5,41,33,29,18,27, 5,14,24,11,51, 6,40
0	7	21	1,43,28,49, 1,47,26,46,38,26,15	34,47,21,14,46,51,53,37,32, 9,36
28	0	1	1,43,33,47, 1,20	34,45,41, 8,37,40,32,48,45
9	6	0	1,43,40,48	34,43,20
0	17	4	1,43,47,49,27,11,15	34,40,59, 0,55,25, 6,10,22,13,20
32	0	12	1,44, 2,57, 2,46,15,18,31, 6,40	34,35,56,29, 3,45
1	0	5	1,44,10	34,33,36
0	15	13	1,44,17, 3,25,53, 1,38,26,15	34,31,15,40,26,40
25	22	0	1,44,29, 7,59,15,34,43,58, 4,48	34,27,16,17,15,34,25,50,37, 2,13,20
8	29	0	1,44,36,12,42,55, 6,31,40,48	34,24,56,23,24, 0,28,56,44,31,50,31,18,22,19,55, 3,42,13,20
5	0	16	1,44,39,20,13,25,33,20	34,23,54,43,42,24,46, 4,48
47	0	3	1,44,44,21,36, 2,41,11, 6,40	34,22,15,45, 0,20,18,58, 2,46,33, 9,50,37,30
24	4	0	1,44,51,27,21,36	34,19,56,11,29, 3,45
7	11	0	1,44,58,33,36	34,17,36,47,24,26,40
0	23	5	1,45, 5,40,19,16,38,26,15	34,15,17,32,45,50,43, 8, 1,12,25,40,44,26,40
20	0	7	1,45,20,59,15,33,20	34,10,18,45
1	6	6	1,45,28, 7,30	34, 8
6	34	0	1,45,54,39,52,27,17,51,34,33,36	33,59,26,48,17,47, 8,35,18, 3,18, 2,46,17,36,42,31,48,22, 3,27,24,26,40
1	4	15	1,45,57,49,43,35,37,30	33,58,25,54,16,42,14,24
39	2	0	1,46, 2,54,52,14,43,12	33,56,48, 8,53,40, 3,55, 6,26,43, 7,30
22	9	0	1,46,10, 5,57, 7,12	33,54,30,18,45
5	16	0	1,46,17,17,31,12	33,52,12,37,55,59,40,14,48,53,20
0	29	6	1,46,24,29,34,31, 5,55, 4,41,15	33,49,55, 6,26, 1,12,13,50,49,18,41,43, 9,18, 1,28,53,20
1	2	24	1,46,27,40,19, 7,34,42,33, 7,30	33,48,54,29,29,13,57,38,38,41,58,16,19,12
39	0	9	1,46,32,46,53,43, 2,42,57,46,40	33,47,17,11,30,22,51, 5,37,30
8	0	2	1,46,40	33,45
1	12	7	1,46,47,13,35,37,30	33,42,42,57,46,40
12	0	13	1,47,10, 2,28, 8,53,20	33,35,32,21, 7,12
54	0	0	1,47,15,11, 4,44,59,27,13, 4	33,33,55,41,36,25,27,53,52,23,53,56,57,74,26, 0,56,15
1	10	16	1,47,17,18, 5,53,19,13, 7,30	33,33,15,57,18,43,12
37	7	0	1,47,22,27, 3,23,54,14,24	33,31,39,24,20,24,45,21, 5,37,30
20	14	0	1,47,29,43,31,35, 2,24	33,29,23,16,17,46,40
3	21	0	1,47,37, 0,29,20,24	33,27, 7,17,27,53,44,56, 6,48,13,49,37,46,40
27	0	4	1,47,52,41,28,53,20	33,22,15,29,52,58, 7,30
2	3	0	1,48	33,20
1	18	8	1,48, 7,19, 0,49,13, 7,30	33,17,44,39,17,12, 5,55,33,20
0	0	8	1,48,30,25	33,10,35,21,36
52	5	0	1,48,35,37,28, 3,33,11,48,28,48	33, 9, 3,53,41, 9,35,42, 5,49,31,48, 6,19,41,15
35	12	0	1,48,42,58,53,41,27,10, 4,48	33, 6,49,17,22,22,58, 7,30
18	19	0	1,48,50,20,49,13,43,40,48	33, 4,34,50,10, 9, 3,12,35,33,20
1	26	0	1,48,57,43,14,42,24,18	33, 2,20,32, 3,50,51,47,16,20,58, 6, 3,14,14,19,15,33,20
4	0	19	1,49, 0,58,33,59, 7,13,20	33, 1,21,20,21,30,58,38,12,28,48
46	0	6	1,49, 6,12,30, 2,47,54, 4,26,40	32,59,46,19,12,19,30,12,31,27,53,26,15
17	1	0	1,49,13,36	32,57,32,20,37,30
0	8	0	1,49,21	32,55,18,31, 6,40
1	24	9	1,49,28,24,30, 4,50, 2,20,37,30	32,53, 4,50,39,12,41,24,29,57,31,51, 6,40
19	0	10	1,49,44,21,43,42,13,20	32,48,18
0	6	9	1,49,51,47,48,45	32,46, 4,48
33	17	0	1,50, 4,31, 7,51,43,15,27,21,36	32,42,17,34,11,44,10
16	24	0	1,50,11,58,34,50,38,58,33,36	32,40, 4,46,35,12,38,43,32,53,39,45,11, 6,40
1	32	1	1,50,19,26,32, 8,26, 6,13,30	32,37,52, 7,57,52,27,26,41,19,58, 7,27,38,30,26,25,44, 1,58,31, 6,40
0	4	18	1,50,22,44,17,54,36,33,45	32,36,53,40, 6,26, 9, 1,26,24
34	0	1	1,50,28, 2, 9,25,20	32,35,19,49,20,19,15,45,42,11,15
15	6	0	1,50,35,31,12	32,33, 7,30
0	14	1	1,50,43, 0,45	32,30,55,19,36,57,17, 2,13,20
7	0	5	1,51, 6,40	32,24
0	12	10	1,51,14,11,39,36,33,45	32,21,48,26,40
14	29	0	1,51,34,37,33,46,46,57,47,31,12	32,15,52,51,56,15,27, 8,11,44,51, 6,50,58,26,10,22,13,20
11	0	16	1,51,37,57,34,19,15,33,20	32,14,55, 3,28,30,43,12
53	0	3	1,51,43,19, 2,26,51,55,51, 6,40	32,13,22,15,56,34, 2,46,55, 6, 8,35,28,42,39,22,30
0	10	19	1,51,45,31,21, 8, 2,31,10,18,45	32,12,44, 7, 1,10,16,19,12
30	4	0	1,51,50,53,11, 2,24	32,11,11,25,45,59,45,56,15
13	11	0	1,51,58,27,50,24	32, 9, 0,44,26,40
0	20	2	1,52, 6, 3, 0,33,45	32, 6,50,11,57,58,47,56,16, 7,54, 4,26,40
26	0	7	1,52,22,23,12,35,33,20	32, 2,10, 4,41,15
1	3	3	1,52,30	32
0	18	11	1,52,37,37,18,21,16,10,18,45	31,57,50, 4, 6,54,48,53,20
1	1	12	1,53, 1,41, 2,30	31,51, 1,47, 8, 9,36
45	2	0	1,53, 7, 6,31,43,42, 4,48	31,49,30, 8,20,18,40,24,47,32,55,46,52,30
28	9	0	1,53,14,46,20,55,40,48	31,47,20,55, 4,41,15
11	16	0	1,53,22,26,41,16,48	31,45,11,50,33,44,41,28,53,20
0	26	3	1,53,30, 7,32,49,10,18,45	31,43, 2,54,46,53,37,42,58,53,43,46,36,42,28, 8,53,20
3	0	22	1,53,33,31, 0,24, 5, 1,23,20	31,42, 6, 5, 8,39,20,17,28,46,50,52,48
14	0	2	1,53,46,40	31,38,26,15
1	9	4	1,53,54,22,30	31,36,17,46,40
18	0	13	1,54,18,42,38, 1,28,53,20	31,29,34, 4,48
60	0	0	1,54,24,11,49, 3,59,25, 1,56,16	31,28, 3,27,45,23,52,24,15,22,24,19,38,49, 9,23,22,44, 3,45
1	7	13	1,54,26,27,18,16,52,30	31,27,26,12,28,48
43	7	0	1,54,31,56,51,37,29,51,21,36	31,25,55,41,34, 8,12,31, 1,31,24,22,30
26	14	0	1,54,39,42,25,41,22,33,36	31,23,48, 4, 1,40
9	21	0	1,54,47,28,31,17,45,36	31,21,40,35, 7,24, 8,22,36,22,42,57,46,40
0	32	4	1,54,55,15, 8,28,47,11,29, 3,45	31,19,33,14,50,45,33,32,49,16,46,11,57,44,10, 1,22,18,16,17,46,40
1	5	22	1,54,58,41, 8,39,23, 5, 9,22,30	31,18,37, 7,18,10,42,15,46,56,38,24
33	0	4	1,55, 4,12,14,48,53,20	31,17, 7, 1,45,54,29,31,52,30
8	3	0	1,55,12	31,15
1	15	5	1,55,19,48,16,52,30	31,12,53, 6,49,52,35,33,20
6	0	8	1,55,44,26,40	31, 6,14,24
1	13	14	1,55,52,17, 8,45,35, 9,22,30	31, 4, 8, 6,24
41	12	0	1,55,57,50,49,16,12,58,45, 7,12	31, 2,38,42,32,14, 1,59,31,52,30
24	19	0	1,56, 5,42,12,30,38,35,31,12	31, 0,32,39,32, 0,59,15,33,20
7	26	0	1,56,13,34, 7,41,13,55,12	30,58,26,45, 3,36,26, 3, 4, 4,39,28,10,32, 5,55,33,20
10	0	19	1,56,17, 2,28,15, 3,42,13,20	30,57,31,15,20,10,17,28,19,12

P	Q	R	
23·	1	0	1,56,30,30,24
6	8	0	1,56,38,24
1	21	6	1,56,46,18, 8, 5, 9,22,30
25	0	10	1,57, 3,19,10,37, 2,13,20
0	3	6	1,57,11,15
22	24	0	1,57,32,46,29,10, 1,34,27,50,24
5	31	0	1,57,40,44,18,16,59,50,38,24
0	1	15	1,57,44,15,15, 6,15
40	0	1	1,57,49,54,18, 3, 1,20
21	6	0	1,57,57,53,16,48
4	13	0	1,58, 5,52,48
1	27	7	1,58,13,52,51,41,13,14,31,52,30
2	0	25	1,58,17,24,47,55, 5,13,56,48,20
13	0	5	1,58,31, 6,40
0	9	7	1,58,39, 8,26,15
17	0	16	1,59, 4,29,24,36,32,35,33,20
3	36	0	1,59, 8,59,51,30,42,35,31,22,48
0	7	16	1,59,12,33,26,32,34,41,15
36	4	0	1,59,18,16,43,46,33,36
19	11	0	1,59,26,21,41,45,36
2	18	0	1,59,34,27,12,36
32	0	7	1,59,51,52,45,25,55,33,20
1	0	0	2
0	15	8	2, 0, 8, 7,47,34,41,15
5	0	11	2, 0,33,47,46,40
51	2	0	2, 0,39,34,57,50,36,53, 7,12
0	13	17	2, 0,41,57,51,37,29, 7,15,56,15
34	9	0	2, 0,47,45,26,19,23,31,12
17	16	0	2, 0,55,56,28, 1,55,12
0	23	0	2, 1, 4, 8, 3, 0,27
9	0	22	2, 1, 7,45, 4,25,41,21,28,53,20
20	0	2	2, 1,21,46,40
1	6	1	2, 1,30
0	21	9	2, 1,38,13,53,25,22,15,56,15
24	0	13	2, 1,55,57,28,33,34,48,53,20
1	4	10	2, 2, 4,13, 7,30
49	7	0	2, 2,10, 4,39, 3,59,50,47, 2,24
32	14	0	2, 2,18,21,15,24, 8, 3,50,24
15	21	0	2, 2,26,38,25,22,56,38,24
0	29	1	2, 2,34,56, 9, 2,42,20,15
1	2	19	2, 2,38,35,53,14, 0,37,30
39	0	4	2, 2,44,29, 3,48, 8,53,20
14	3	0	2, 2,52,48
1	12	2	2, 3, 1, 7,30
12	0	8	2, 3,27,24,26,40
1	10	11	2, 3,35,46,17,20,37,30
30	19	0	2, 3,50, 5, 1,20,41, 9,53,16,48
13	26	0	2, 3,58,28,24,11,58,5C,52,48
16	0	19	2, 4, 2,10,38, 8, 3,57, 2,13,20
0	35	2	2, 4, 6,52,21, 9,29,22, 0,11,15
1	8	20	2, 4,10,34,50, 8,56, 7,58, 7,30
29	1	0	2, 4,16,32,25,36
12	8	0	2, 4,24,57,36
1	18	3	2, 4,33,23,20,37,30
31	0	10	2, 4,51,32,27,19,30,22,13,20
0	0	3	2, 5
1	16	12	2, 5, 8,28, 7, 3,37,58, 7,30
11	31	0	2, 5,31,27,15,30, 7,50, 0,57,36
4	0	14	2, 5,35,12,16, 6,40
46	0	1	2, 5,41,13,55,15,13,25,20
27	6	0	2, 5,49,44,49,55,12
10	13	0	2, 5,58,16,19,12
1	24	4	2, 6, 6,48,23, 7,58, 7,30
19	5	0	2, 6,25,11, 6,40
0	6	4	2, 6,33,45
23	0	16	2, 7, 0,47,22,14,58,45,55,33,20
0	4	13	2, 7, 9,23,40,18,45
42	4	0	2, 7,15,29,50,41,39,50,24
25	11	0	2, 7,24, 7, 8,32,38,24
8	18	0	2, 7,32,45, 1,26,24
1	30	5	2, 7,41,23,29,25,19, 6, 5,37,30
0	2	22	2, 7,45,12,22,57, 5,39, 3,45
38	0	7	2, 7,51,20,16,27,39,15,33,20
7	0	0	2, 8
0	12	5	2, 8, 8,40,18,45
11	0	11	2, 8,36, 2,57,46,40
57	2	0	2, 8,42,13,17,41,59,20,39,40,48
0	10	14	2, 8,44,45,43, 3,59, 3,45
40	9	0	2, 8,50,56,28, 4,41, 5,16,48
23	16	0	2, 8,59,40,13,54, 2,52,48
6	23	0	2, 9, 8,24,35,12,28,48
26	0	2	2, 9,27,13,46,40
5	5	0	2, 9,36
0	18	6	2, 9,44,46,48,59, 3,45
30	13	0	2,10, 3,41,18,27,49, 8, 8,53,20
1	1	7	2,10,12,30
0	16	15	2,10,21,19,17,21,17, 3, 2,48,45
38	14	0	2,10,27,34,40,25,44,36, 5,45,36
21	21	0	2,10,36,24,59, 4,28,24,57,36
4	28	0	2,10,45,15,53,38,53, 9,36
3	0	17	2,10,49,10,16,46,56,40
45	0	4	2,10,55,27, 0, 3,21,28,53,20
20	3	0	2,11, 4,19,12
3	10	0	2,11,13,12
0	24	7	2,11,22, 5,24, 5,48, 2,48,45
18	0	8	2,11,41,14, 4,26,40
1	7	8	2,11,50, 9,22,30
19	26	0	2,12,14,22,17,48,46,46,16,19,12
2	33	0	2,12,23,19,50,34, 7,19,28,12
1	5	17	2,12,27,17, 9,29,31,52,30

30,53,56,34,2C, 9,22,30
30,51,51, 6,40
30,49,45,47,29,15,38,49,13, 5,11, 6,40
30,45,16,52,30
30,43,12
30,37,34,28,40,30,36,18,19,35,18,31, 6,40
30,35,3C, 7,28, 0,25,43,46,14,58,14,29,39,51, 2,16,37,31,51, 6,40
30,34,35,18,51, 2, 0,57,36
30,33, 7,20, 0,18, 3,31,35,48, 2,48,45
30,31, 3,16,52,30
30,28,55,22, 8,23,42,13,20
30,26,55,35,47,25, 5, 0,27,44,22,49,32,50,22,13,20
30,26, 1, 2,32,18,33,52,46,49,46,26,41,16,48
30,22,3C
30,20,26,40
30,13,59, 7, 0,28,48
30,12,5C,29,35,48,34,18, 2,42,56, 2,27,48,59,17,48,16,19,36,24,21,43,42,13,20
30,11,56,21,34,50,52,48
30,10,29,27,54,22,16,48,59, 3,45
30, 8,26,56,40
30, 6,24,33,43, 6,22,26,3C, 7,24,26,40
30, 2, 1,56,53,40,18,45
30
29,57,56,11,21,28,53,20
29,51,35,25,26,24
29,50, 5,30,19, 2,38, 7,53,14,34,37,17,41,43, 7,30
29,49,34,10,56,38,24
29,48, 8,21,38, 8,4C,18,45
29,46, 7,21, 9, 8, 8,53,2C
29,44, 6,28,51,27,46,36,32,42,52,17,26,54,48,53,20
29,39,47, 6,33,45
29,37,46,40
29,35,46,21,35,17,25,16, 2,57,46,40
29,31,28,12
29,29,28,19,12
29,28, 3,27,43,15,11,44, 5,10,41,36, 5,37,30
29,26, 3,48,46,33,45
29,24, 4,17,55,41,22,51,11,36,17,46,40
29,22, 4,55,10, 5,12,42, 1,11,58,18,42,52,39,23,47, 9,37,46,40
29,21,12,18, 5,47,32, 7,17,45,36
29,19,47,50,24,17,2C,11, 7,58, 7,30
29,17,48,45
29,15,49,47,39,15,33,20
29, 9,36
29, 7,37,36
29, 4,15,37, 3,45,55,33,20
29, 2,17,34,44,37,34,24,22, 0, 9,52,35,33,20
29, 1,25,33, 7,39,38,52,48
29, 0,19,40,24,46,37,43,43,24,24,59,57,54,13,43,29,32,28,25,20,59,15,33,20
28,59,27,42,19, 3,14,41,16,48
28,58, 4,17,11,23,47,20,37,30
28,56, 6,40
28,54, 9,10,46,10,55, 8,38,31, 6,40
28,49,57, 4,13, 7,30
28,48
28,46, 2, 3,42,13,20
28,40,46,55,56,24, 0,24, 7,17, 6,32, 6, 5,18,36,35,53, 5,11, 6,40
28,39,55,36,25,20,38,24
28,38,33, 7,3C,16,55,48,22,18,47,38,12,11,15
28,36,36,49,34,13, 7,30
28,34,40,39,30,22,13,20
28,32,44,37,18,12,15,56,41, 0,21,23,57, 2,13,20
28,28,35,37,30
28,26,40
28,20,36,40,19,12
28,18,41,35,13,55,12
28,17,20, 7,24,43,23,15,55,22,15,56,15
28,15,25,15,37,30
28,13,30,31,36,39,43,32,2C,44,26,40
28,11,35,55,21,41, 0,11,32,21, 5,34,45,57,45, 1,14, 4,26,40
28,10,45,24,34,21,38, 2,12,14,58,33,36
28, 9,24,19,35,19, 2,34,41,15
28, 7,30
28, 5,35,48, 8,53,2C
27,59,36,57,36
27,58,16,24,40,21,13,14,53,39,54,57,27,50,21,40,46,52,30
27,57,43,17,45,36
27,56,22,50,17, 0,37,47,34,41,15
27,52,36, 4,33,14,47,26,45,40,11,31,21,28,53,20
27,48,32,54,54, 8,26,15
27,46,4C
27,44,47,12,44,20, 4,56,17,46,40
27,40,45,11,15
27,38,52,48
27,37, C,32,21,20
27,35,41, 4,28,39, 8,26,15
27,33,49, 1,48,27,32,40,29,37,46,40
27,31,57, 6,43,12,23, 9,23,37,28,25, 2,41,51,56, 2,57,46,40
27,31, 7,46,57,55,48,51,50,24
27,29,48,36, 0,16,15,10,26,13,14,31,52,30
27,27,56,57,11,15
27,26, 5,25,55,33,20
27,24,14, 2,12,40,34,30,24,57,56,32,35,33,20
27,20,15
27,18,24
27,13,23,58,49,20,32,16,17,24,43, 7,39,15,33,20
27,11,33,26,38,13,42,52,14,26,38,26,13, 2, 5,22, 1,26,41,38,45,55,33,20
27,10,44,43,25,21,47,31,12

P	Q	R	
35	1	0	2,12,33,38,35,18,24
18	8	0	2,12,42,37,26,24
1	15	0	2,12,51,36,54
37	0	10	2,13,10,58,37, 8,48,23,42,13,20
6	0	3	2,13,20
1	13	9	2,13,29, 1,59,31,52,30
10	0	14	2,13,57,33, 5,11, 6,40
0	38	0	2,14, 2,37,20,27, 2,54,57,48, 9
52	0	1	2,14, 3,58,50,56,14,19, 1,20
1	11	18	2,14, 6,37,37,21,39, 1,24,22,30
33	6	0	2,14,13, 3,49,14,52,48
16	13	0	2,14,22, 9,24,28,48
1	21	1	2,14,31,15,36,40,30
25	0	5	2,14,50,51,51, 6,40
0	3	1	2,15
1	19	10	2,15, 9, 8,46, 1,31,24,22,30
0	1	10	2,15,38, 1,15
48	4	0	2,15,44,31,50, 4,26,29,45,36
31	11	0	2,15,53,43,37, 6,48,57,36
14	18	0	2,16, 2,56, 1,32, 9,36
1	27	2	2,16,12, 9, 3,23, 0,22,30
2	0	20	2,16,16,13,12,28,54, 1,40
44	0	7	2,16,22,45,37,33,29,52,35,33,20
13	0	0	2,16,32
0	9	2	2,16,41,15
17	0	11	2,17,10,27, 9,37,46,40
0	7	11	2,17,19,44,45,56,15
46	9	0	2,17,26,20,13,56,59,49,37,55,12
29	16	0	2,17,35,38,54,49,39, 4,19,12
12	23	0	2,17,44,58,13,33,18,43,12
1	33	3	2,17,54,18,10,10,32,37,46,52,30
0	5	20	2,17,58,25,22,23,15,42,11,15
32	0	2	2,18, 5, 2,41,46,40
11	5	0	2,18,14,24
0	15	3	2,18,23,45,56,15
5	0	6	2,18,53,20
0	13	12	2,19, 2,44,34,30,42,11,15
27	21	0	2,19,18,50,39, 0,46,18,37,26,24
10	28	0	2,19,28,16,57,13,28,42,14,24
9	0	17	2,19,32,26,57,54, 4,26,40
51	0	4	2,19,39, 8,48, 3,34,54,48,53,20
26	3	0	2,19,48,36,28,48
9	10	0	2,19,58, 4,48
0	21	4	2,20, 7,33,45,42,11,15
24	0	8	2,20,27,59, 0,44,26,40
1	4	5	2,20,37,30
0	19	13	2,20,47, 1,37,56,35,12,53,26,15
8	33	0	2,21,12,53, 9,56,23,48,46, 4,48
1	2	14	2,21,17, 6,18, 7,30
41	1	0	2,21,23,53, 9,39,37,36
24	8	0	2,21,33,27,56, 9,36
7	15	0	2,21,43, 3,21,36
0	27	5	2,21,52,39,26, 1,27,53,26,15
1	0	23	2,21,56,53,45,30, 6,16,44,10
12	0	3	2,22,13,20
1	10	6	2,22,22,58, 7,30
16	0	14	2,22,53,23,17,31,51, 6,40
58	0	1	2,23, 0,14,46,19,59,16,17,25,20
1	8	15	2,23, 3, 4, 7,51, 5,37,30
39	6	0	2,23, 9,56, 4,31,52,19,12
22	13	0	2,23,19,38, 2, 6,43,12
5	20	0	2,23,29,20,39, 7,12
31	0	5	2,23,50,15,18,31, 6,40
4	2	0	2,24
1	16	7	2,24, 9,45,21, 5,37,30
4	0	9	2,24,40,33,20
54	4	0	2,24,47,29,57,24,44,15,44,38,24
1	14	16	2,24,50,21,25,56,58,56,43, 7,3C
37	11	0	2,24,57,18,31,35,16,13,26,24
20	18	0	2,25, 7, 7,45,38,18,14,24
3	25	0	2,25,16,57,39,36,32,24
8	0	20	2,25,21,18, 5,18,49,37,46,40
19	0	0	2,25,38, 8
2	7	0	2,25,48
1	22	8	2,25,57,52,40, 6,26,43, 7,30
23	0	11	2,26,19, 8,58,16,17,46,40
0	4	8	2,26,29, 3,45
35	16	0	2,26,46, 1,30,28,57,40,36,28,48
18	23	0	2,26,55,58, 6,27,31,58, 4,48
1	30	0	2,27, 5,55,22,51,14,48,18
0	2	17	2,27,10,19, 3,52,48,45
38	0	2	2,27,17,22,52,33,46,40
17	5	0	2,27,27,21,36
0	12	0	2,27,37,21
0	0	26	2,27,51,45,59,53,51,32,26, 0,25
11	0	6	2,28, 8,53,20
0	10	9	2,28,18,55,32,48,45
16	28	0	2,28,46,10, 5, 2,22,37, 3,21,36
15	0	17	2,28,50,36,45,45,40,44,26,40
1	36	1	2,28,56,14,49,23,23,14,24,13,2C
0	8	18	2,29, 0,41,48,10,43,21,33,45
32	3	0	2,29, 7,50,54,43,12
15	10	0	2,29,17,57, 7,12
0	18	1	2,29,28, 4, 0,45
30	0	8	2,29,49,50,56,47,24,26,40
1	1	2	2,30
0	16	10	2,30,10, 9,44,28,21,33,45
3	0	12	2,30,42,14,43,20
47	1	0	2,30,49,28,42,18,16, 6,24
30	8	0	2,30,59,41,47,54,14,24

27, 9,26,31, 6,56, 3, 8, 5, 9,22,30
27, 7,36,15
27, 5,46, 6,20,47,44,11,51, 6,40
27, 1,49,45,12,18,16,52,30
27
26,58,1C,22,13,20
26,52,25,52,53,45,36
26,51,24,52,58,29,50,29,22,24,49,48,51,?3,32,42,29,34,30,45,41,39,18,50,51,51, 6,40
26,51, 8,33,17, 8,22,19, 5,55, 7, 9,33,55,32,48,45
26,50,36,45,50,58,33,36
26,49,19,31,28,19,48,16,52,30
26,47,3C,37, 2,13,20
26,45,41,49,58,18,59,56,53,26,35, 3,42,13,20
26,41,48,23,54,22,30
26,4C
26,38,11,43,25,45,40,44,26,40
26,32,31,29,16,48
26,31,15, 6,56,55,40,33,40,39,37,26,29, 3,45
26,29,27,25,53,54,22,30
26,27,3S,52, 8, 7,14,34, 4,26,40
26,25,52,25,39, 4,41,25,49, 4,46,28,50,35,23,27,24,26,40
26,25, 5, 4,17,12,46,54,33,59, 2,24
26,23,4S, 3,21,51,36,10, 1,10,18,45
26,22, 1,52,30
26,2C,14,48,53,2C
26,14,38,24
26,12,51,50,24
26,11,36,24,38,26,50,25,51,16,10,18,45
26, 9,5C, 3,21,23,20
26, 8, 3,49,16,1C, 6,58,50,18,55,48, 8,?3,20
26, 6,17,42,22,17,57,57,21, 3,58,29,58, 6,48,21, 8,35,13,34,48,53,20
26, 5,30,56, 5, 8,55,13, 9, 7,12
26, 4,15,51,28,15,24,36,33,45
26, 2,3C
26, 0,44,15,41,33,49,37,46,40
25,55,12
25,53,26,45,20
25,50,27,12,56,40,49,22,57,46,40
25,48,42,17,33, 0,21,42,33,23,52,53,28,46,44,56,17,46,40
25,47,56, 2,46,48,34,33,36
25,46,41,48,45,15,14,13,32, 4,54,52,22,58, 7,30
25,44,57, 8,36,47,48,45
25,43,12,35,33,20
25,41,28, 9,34,23, 2,21, 0,54,19,15,33,?0
25,37,44, 3,45
25,36
25,34,16, 3,17,31,51, 6,4C
25,29,35, 6,13,20,21,26,28,32,28,32, 4,43,12,31,53,51,16,32,35,33,20
25,28,49,25,42,31,40,48
25,27,36, 6,40,15, 2,56,19,50, 2,20,37,30
25,25,52,44, 3,45
25,24, 9,28,26,59,45,11, 6,40
25,22,26,19,49,30,54,10,23, 6,59, 1,17,21,58,31, 6,40
25,21,4C,52, 6,55,28,13,59, 1,28,42,14,?4
25,18,45
25,17, 2,13,20
25,11,3S,15,5C,24
25,10,26,46,12,19, 5,55,24,17,55,27,43, 3,19,30,42,11,15
25, 9,5C,57,59, 2,24
25, 8,44,33,11,18,34, 0,49,13, 7,30
25, 7, 2,27,13,20
25, 5,2C,28, 5,59,18,42, 5, 6,10,22,13,?0
25, 1,41,37,24,43,35,37,30
25
24,58,18,29,27,54, 4,26,40
24,52,56,31,12
24,51,47,55,15,52,11,46,34,22, 8,51, 4,44,45,56,15
24,51,18,29, 7,12
24,50, 6,58, 1,47,13,35,37,30
24,48,26, 7,37,36,47,24,26,40
24,46,45,24, 2,53, 8,50,27,15,43,34,32,25,40,44,26,40
24,46, 1, 0,16, 8,13,58,39,21,36
24,43, 9,15,28, 7,3C
24,41,28,53,2C
24,39,48,37,59,24,31, 3,22,28, 8,53,20
24,36,13,30
24,34,33,36
24,31,43,10,38,48, 7,30
24,30, 3,34,56,24,29, 2,39,40,14,48,53,20
24,28,24, 5,58,24,2C,35, 0,59,58,35,35,43,52,49,49,18, 1,28,53,20
24,27,40,15, 4,49,36, 4,48
24,26,25,52, 0,14,26,49,16,38,26,15
24,24,50,37,30
24,23,11,29,42,42,57,46,40
24,20,48,50, 1,50,51, 6,13,27,49, 9,21, 1,26,24
24,18
24,16,21,20
24,11,54,38,57,11,35,21, 8,48,38,20, 8,13,49,37,46,40
24,11,11,17,36,23, 2,24
24,10,1C,23,4C,38,51,26,26,1C,20,49,58,15,11,26,14,37, 3,41, 7,29,22,57,46,40
24, 9,33, 5,15,52,42,14,24
24, 8,23,34,19,29,49,27,11,15
24, 6,45,33,2C
24, 5, 7,38,58,29, 5,57,12, 5,55,33,20
24, 1,37,33,30,56,15
24
23,58,22,33, 5,11, 6,40
23,53,16,20,21, 7,12
23,52, 7,36,15,14, 6,30,18,35,39,41,50, 9,22,30
23,50,3C,41,18,30,56,15

```
 P  Q  R
13 15  0    2,31, 9,55,35, 2,24                        23,48,53,52,55,18,31, 6,40
 0 24  2    2,31,20,10, 3,45,33,45                      23,47,17,11, 5,10,13,17,14,10,17,49,57,31,51, 6,40
 7  0 23    2,31,24,41,20,32, 6,41,51, 6,4C             23,46,24,33,51,29,30,13, 6,35, 8, 9,36
18  0  3    2,31,42,13,20                               23,43,49,41,15
 1  7  3    2,31,52,30                                  23,42,13,20
 0 22 11    2,32, 2,47,21,46,42,49,55,18,45             23,40,37, 5,16,13,56,12,50,22,13,20
22  0 14    2,32,24,56,50,41,58,31, 6,40                23,37,1C,23,36
 1  5 12    2,32,35,16,24,22,30                         23,35,34,39,21,36
45  6  0    2,32,42,35,48,49,59,48,28,48                23,34,26,46,10,36, 9,23,16, 8,33,16,52,30
28 13  0    2,32,52,56,34,15,10, 4,48                   23,32,51, 3, 1,15
11 20  0    2,33, 3,18, 1,43,40,48                      23,31,15,26,20,33, 6,16,57,17, 2,13,20
 0 30  3    2,33,13,40,11,18,22,55,18,45                23,29,39,56, 8, 4,1C, 9,36,57,34,38,58,18, 7,31, 1,43,42,13,20
 1  3 21    2,33,18,14,51,32,30,46,52,30                23,28,57,50,28,38, 1,41,5C,12,28,48
37  C  5    2,33,25,36,19,45,11, 6,40                   23,27,5C,16,19,25,52, 8,54,22,30
10  2  0    2,33,36                                     23,26,15
 1 13  4    2,33,46,24,22,30                            23,24,39,50, 7,24,26,40
10  0  9    2,34,19,15,33,20                            23,19,4C,48
 1 11 13    2,34,29,42,51,40,46,52,30                   23,18, 6, 4,48
43 11  0    2,34,37, 7,45,41,37,18,20, 9,36             23,16,59, 1,54,10,31,29,38,54,22,30
26 18  0    2,34,47,36,16,40,51,27,21,36                23,15,24,29,39, 0,44,26,40
 9 25  0    2,34,58, 5,30,14,58,33,36                   23,13,5C, 3,47,42,19,32,18, 3,29,36, 7,54, 4,26,40
14  0 20    2,35, 2,43,17,40, 4,56,17,46,4C             23,13, 8,26,30, 7,43, 6,14,24
25  0  0    2,35,20,40,32                               23,10,27,25,45, 7, 1,52,30
 8  7  0    2,35,31,12                                  23, 8,53,20
 1 19  5    2,35,41,44,10,46,52,30                      23, 7,15,20,36,56,44, 6,54,48,53,20
29  0 11    2,36, 4,25,34, 9,22,57,46,40                23, 3,57,39,22,30
 0  1  5    2,36,15                                     23, 2,24
 1 17 14    2,36,25,35, 8,49,32,27,39,22,20             23, C,5C,26,57,46,40
24 23  0    2,36,43,41,58,53,22, 5,57, 7,12             22,58,1C,51,30,22,57,13,44,41,28,53,20
 7 30  0    2,36,54,19, 4,22,39,47,31,12                22,56,37,35,36, 0,19,17,49,41,13,40,52,14,53,16,42,28, 8,53,20
 2  0 15    2,36,59, 0,20, 8,20                         22,55,5C,29, 8,16,3C,43,12
44  0  2    2,37, 6,32,24, 4, 1,46,40                   22,54,5C,30, 0,13,32,38,41,51, 2, 6,33,45
23  5  0    2,37,17,11, 2,24                            22,53,17,27,39,22,30
 6 12  0    2,37,27,50,24                               22,51,44,31,36,17,46,40
 1 25  6    2,37,38,30,28,54,57,39,22,30                22,50,11,41,50,33,48,45,2C,48,17, 7, 9,37,46,40
17  0  6    2,38, 1,28,53,20                            22,46,52,30
 0  7  6    2,38,12,11,15                               22,45,2C
21  0 17    2,38,45,59,12,48,43,27,24,26,4C             22,40,25,20,15,21,36
 5 35  0    2,38,51,59,48,40,56,47,21,50,24             22,39,37,52,11,51,25,43,32, 2,12, 1,50,51,44,28,21,12,14,42,18,16,17,46,40
 0 15  0    2,38,56,44,35,23,26,15                      22,38,57,16,11, 8, 9,36
38  3  0    2,39, 4,22,18,22, 4,48                      22,37,52, 5,55,46,42,36,44,17,48,45
21 10  0    2,39,15, 8,55,40,48                         22,36,2C,12,30
 4 17  0    2,39,25,56,16,48                            22,34,48,25,17,19,46,49,52,35,33,20
 0  3 24    2,39,41,30,28,41,22, 3,49,41,15             22,32,36,19,39,29,18,25,45,47,58,50,52,48
36  0  8    2,39,49,10,20,34,34, 4,26,40                22,31,31,27,40,15,14, 3,45
 5  0  1    2,40                                        22,30
 0 13  7    2,40,10,50,23,26,15                         22,28,28,38,31, 6,40
 9  0 12    2,40,45, 3,42,13,20                         22,23,41,34, 4,48
53  1  0    2,40,52,46,37, 7,29,10,49,36                22,22,37, 7,44,16,58,35,54,55,55,57,58,16,17,20,37,30
 0 11 16    2,40,55,57, 8,49,58,49,41,15                22,22,1C,38,12,28,48
36  8  0    2,41, 3,40,35, 5,51,21,36                   22,21, 6,16,13,36,30,14, 3,45
19 15  0    2,41,14,35,17,22,33,36                      22,19,35,30,51,51, 6,40
 2 22  0    2,41,25,30,44, 0,36                         22,18, 4,51,38,35,49,57,24,32, 9,13, 5,11, 6,40
24  0  3    2,41,49, 2,13,20                            22,14,5C,19,55,18,45
 1  4  0    2,42                                        22,13,20
 0 19  8    2,42,10,58,31,13,49,41,15                   22,11,49,46,11,28, 3,57, 2,13,20
28  0 14    2,42,34,36,38, 4,46,25,11, 6,40             22, 8,3C, 9
 1  2  9    2,42,45,37,30                               22, 7, 6,14,24
51  6  0    2,42,53,26,12, 5,19,47,42,43,12             22, 6, 2,35,47,26,23,48, 3,53, 1,12, 4,13, 7,30
34 13  0    2,43, 4,28,20,32,10,45, 7,12                22, 4,32,51,34,55,18,45
17 2C  0    2,43,15,31,13,50,35,31,12                   22, 3, 3,13,26,46, 2, 8,23,42,13,20
 0 27  0    2,43,26,34,52, 3,36,27                      22, 1,33,41,22,33,54,31,3C,53,58,44, 2, 9,29,32,50,22,13,20
 1  0 18    2,43,31,27,50,58,40,50                      22, 0,54,13,34,20,39, 5,28,19,12
43  0  5    2,43,39,18,45, 4,11,51, 6,40                21,59,5C,52,48,13, 0, 8,2C,58,35,37,30
16  2  0    2,43,50,24                                  21,58,21,33,45
 1 1C  1    2,44, 1,30                                  21,56,52,20,44,26,40
 0 25  9    2,44,12,36,45, 7,15, 3,30,56,15             21,55,23,13,46, 8,27,36,19,58,21,14, 4,26,40
16  0  9    2,44,36,32,35,33,20                         21,52,12
 1  8 10    2,44,47,41,43, 7,30                         21,50,43,12
32 18  0    2,45, 6,46,41,47,34,53,11, 2,24             21,48,11,42,47,49,26,40
15 25  0    2,45,17,57,52,15,58,27,50,24                21,46,43,11, 3,28,25,49, 1,55,46,30, 7,24,26,40
 0 33  1    2,45,29, 9,48,12,39, 9,20,15                21,45,14,45,18,34,58,17,47,33,18,44,58,25,40,17,37, 9,21,19, 0,44,26,40
 1  6 19    2,45,34, 6,26,51,54,50,37,30                21,45,39,46,44,17,26, 0,57,36
31  C  0    2,45,42, 3,14, 8                            21,43,33,12,53,32,5C,30,28, 7,30
14  7  0    2,45,53,16,48                               21,42, 5
 1 16  2    2,46, 4,31, 7,30                            21,40,36,53, 4,38,11,21,28,53,20
35  0 11    2,46,28,43,16,26, 0,29,37,46,40             21,37,27,48, 9,50,37,30
 4  0  4    2,46,40                                     21,36
 1 14 11    2,46,51,17,29,24,50,37,30                   21,34,32,17,46,40
13 30  0    2,47,21,56,20,40,10,26,41,16,48             21,30,35,14,37,30,18, 5,27,49,54, 4,33,58,57,26,54,48,53,20
 8  0 15    2,47,26,56,21,28,53,20                      21,29,56,42,18,48
50  0  2    2,47,34,58,33,40,17,53,46,40                21,28,54,50,37,42,41,51,16,44, 5,43,39, 8,26,15
29  5  0    2,47,46,19,46,33,36                         21,27,27,37,10,39,50,37,3C
12 12  0    2,47,57,41,45,36                            21,26, 0,29,37,46,40
 1 22  3    2,48, 9, 4,30,50,37,30                      21,24,33,27,58,39,11,57,30,45,16, 2,57,46,40
23  0  6    2,48,33,34,48,53,20                         21,21,26,43, 7,30
 0  4  3    2,48,45                                     21,20
 1 20 12    2,48,56,25,57,31,54,15,28, 7,30             21,18,33,22,44,36,32,35,33,20
 0  2 12    2,49,32,31,33,45                            21,14, 1,11,25,26,24
44  3  0    2,49,40,39,47,35,33, 7,12                   21,13, Cc, 5,33,32,32,26,56,31,41,57,11,15
27 10  0    2,49,52, 9,31,23,31,12                      21,11,33,56,43, 7,30
10 17  0    2,50, 3,40, 1,55,12                         21,10, 7,53,42,29,47,39,15,33,20
 1 28  4    2,50,15,11,19,13,45,28, 7,30                21, 8,41,56,31,15,45, 8,39,15,49,11, 4,78,18,45,55,33,20
 0  0 21    2,50,20,16,30,36, 7,32, 5                   21, 8, 4, 3,25,46,13,31,39,11,13,55,12
42  0  8    2,50,28,27, 1,56,52,20,44,26,40             21, 7, 3,14,41,29,16,56, C,56,15
11  0  1    2,50,40                                     21, 5,37,30
 0 10  4    2,50,51,33,45                               21, 4,11,51, 6,40
15  0 12    2,51,28, 3,57, 2,13,20                      20,59,42,43,12
59  1  0    2,51,36,17,43,35,59, 7,32,54,24             20,58,42,18,30,15,54,56,10,14,56,13, 5,52,46,15,35, 9,22,30
 0  8 13    2,51,39,40,57,25,18,45                      20,58,17,28,19,12
```

P	Q	R		
42	8	0	2,51,47,55,17,26,14,47, 2,24	20,57,17, 7,42,45,28,20,41, 0,56,15
25	15	0	2,51,59,33,38,32, 3,50,24	20,55,52, 2,41, 6,40
8	22	0	2,52,11,12,46,56,38,24	20,54,27, 3,24,56, 5,35, 4,15, 8,38,31, 6,40
0	6	22	2,62,28, 1,42,59, 4,37,44, 3,45	20,52,24,44,52, 7, 8,10,31,17,45,36
30	0	3	2,52,36,18,22,13,20	20,51,24,41,10,36,19,41,15
7	4	0	2,52,48	20,50
0	16	5	2,52,59,42,25,18,45	20,48,35,24,33,15, 3,42,13,20
3	0	7	2,53,36,40	2C,44, 9,36
0	14	14	2,53,48,25,43, 8,22,44, 3,45	20,42,45,24,16
40	13	0	2,53,56,46,13,54,19,28, 7,40,48	2C,41,45,48,21,29,21,19,41,15
23	20	0	2,54, 8,33,18,45,57,53,16,48	20,40,21,46,21,20,39,30,22,13,20
6	27	0	2,54,20,21,11,31,50,52,48	20,38,57,50, 2,24,17,22, 2,43, 6,18,47, 1,23,57, 2,13,20.
7	0	18	2,54,25,33,42,22,35,33,20	20,38, 2C,50,13,26,51,38,52,48
49	0	5	2,54,33,56, 0, 4,28,38,31, 6,40	20,37,21,27, 0,12,11,22,49,39,55,53,54,22,30
22	2	0	2,54,45,45,36	20,35,57,42,53,26,15
5	9	0	2,54,57,36	20,34,34, 4,26,40
0	22	6	2,55, 9,27,12, 7,44, 3,45	20,33,10,31,39,30,25,52,48,43,27,24,26,40
22	0	9	2,55,34,58,45,55,33,20	2C,30,11,15
1	5	7	2,55,46,52,30	20,28,48
21	25	0	2,56,19, 9,43,45, 2,21,41,45,3€	20,25, 2,59, 7, 0,24,12,13, 3,32,20,44,76,40
4	32	0	2,56,31, 6,27,25,29,45,57,36	20,23,40, 4,58,40,17, 9,10,49,58,49,39,46,34, 1,31, 5, 1,14, 4,26,40
1	3	16	2,56,36,22,52,39,22,30	20,23, 3,32,34, 1,20,38,24
37	0	0	2,56,44,51,27, 4,32	20,22, 4,53,20,12, 2,21, 3,52, 1,52,30
20	7	0	2,56,56,49,55,12	20,20,42,11,15
3	14	0	2,57, 8,49,12	20,19,19,34,45,35,48, 8,53,20
0	28	7	2,57,20,49,17,31,49,51,47,48,45	2C,17,57, 3,51,36,43,20,18,29,35,13, 1,53,34,48,53,20
1	1	25	2,57,26, 7,11,52,37,50,55,12,30	20,17,20,41,41,32,22,35,11,13,10,57,47,31,12
10	0	4	2,57,46,40	20,15
1	11	8	2,57,58,42,39,22,30	20,13,37,46,40
14	0	15	2,58,36,44, 6,54,48,53,20	20, 9,15,24,40,19,12
2	37	0	2,58,43,29,47,16, 3,53,17, 4,12	20, 8,33,39,43,52,22,52, 1,48,37,21,38,12,39,31,52,10,53, 4,16,14,29, 8, 8,53,20
56	0	2	2,58,45,18,27,54,59, 5,21,46,40	20, 8,21,24,57,51,16,44,19,26,20,22,10,26,39,36,33,45
1	9	17	2,58,48,50, 9,48,52, 1,52,30	20, 7,57,34,23,13,55,12
35	5	0	2,58,57,25, 5,39,50,24	20, 6,59,38,36,14,51,12,39,22,30
18	12	0	2,59, 9,32,32,38,24	20, 5,37,57,46,40
1	19	0	2,59,21,40,48,54	2C, 4,16,22,28,44,14,57,40, 4,56,17,46,40
29	0	6	2,59,47,49, 8, 8,53,20	20, 1,21,17,55,46,52,30
0	1	0	3	2C
1	17	9	3, 0,12,11,41,22, 1,52,30	19,58,38,47,34,19,15,33,2C
2	0	10	3, 0,50,41,40	19,54,23,36,57,36
50	3	0	3, 0,59,22,26,45,55,19,40,48	19,53,26,20,12,41,45,25,15,29,43, 4,51,47,48,45
33	10	0	3, 1,11,38, 9,29, 5,16,48	19,52, 5,34,25,25,46,52,30
16	17	0	3, 1,23,54,42, 2,52,48	19,50,44,54, 6, 5,25,55,33,20
1	25	1	3, 1,36,12, 4,30,40,30	19,49,24,19,14,18,31, 4,21,48,34,51,37,56,32,35,33,20
6	0	21	3, 1,41,37,36,38,32, 2,13,20	19,48,48,48,12,54,35,10,55,29,16,48
17	0	1	3, 2, 2,40	19,46,31,24,22,30
0	7	1	3, 2,15	19,45,11, 6,40
1	23	10	3, 2,27,20,50, 8, 3,23,54,22,3C	19,43,5C,54,23,31,36,50,41,58,31, 6,40
21	0	12	3, 2,53,56,12,50,22,13,20	19,40,58,48
0	5	10	3, 3, 6,19,41,15	19,39,3€,52,48
48	8	0	3, 3,15, 6,58,35,59,46,10,23,3€	19,38,42,18,28,50, 7,49,23,27, 7,44, 3,45
31	15	0	3, 3,27,31,53, 6,12, 5,45,36	19,37,22,32,31, 2,3C
14	22	0	3, 3,39,57,38, 4,24,57,36	19,36, 2,51,57, 7,35,14, 7,44,11,51, 6,40
1	31	2	3, 3,52,24,13,34, 3,30,22,30	19,34,43,16,46,43,28,28, 0,47,58,52,28,35, 6,15,51,26,25,11, 6,40
0	3	19	3, 3,57,53,49,51, 0,56,15	19,34, 8,12, 3,51,41,24,51,50,24
36	0	3	3, 4, 6,43,35,42,13,20	19,33,11,53,36,11,33,27,25,18,45
13	4	0	3, 4,19,12	19,31,52,30
0	13	2	3, 4,31,41,15	19,30,33,11,46,10,22,13,2C
9	0	7	3, 5,11, 6,40	19,26,24
0	11	11	3, 5,23,39,26, 0,56,15	19,25, 5, 4
29	20	0	3, 5,45, 7,32, 1, 1,44,49,55,12	19,22,5C,24,42,30,37, 2,13,20
12	27	0	3, 5,57,42,36,17,58,16,19,12	19,21,31,43, 9,45,16,16,55, 2,54,40, 6,35, 3,42,13,20
13	0	18	3, 6, 3,15,57,12, 5,55,33,20	19,20,57, 2, 5, 6,25,55,12
0	9	20	3, 6,15,52,15,13,24,11,57,11,15	19,19,38,28,12,42, 4,47,31,12
28	2	0	3, 6,24,48,38,24	19,18,42,51,27,35,51,33,45
11	9	0	3, 6,37,26,24	17,17,24,26,4C
0	19	3	3, 6,50, 5, 0,56,15	19,16, 6, 7,10,47,16,45,45,40,44,26,40
28	0	9	3, 7,17,18,40,59,15,33,20	19,13,18, 2,48,45
1	2	4	3, 7,30	19,12
0	17	12	3, 7,42,42,10,35,26,57,11,15	19,10,42, 2,28, 8,53,20
10	32	0	3, 8,17,10,53,15,11,45, 1,26,24	19, 7,11,19,40, 0,16, 4,51,24,21,24, 3,32,24,23,55,23,27,24,26,40
1	0	13	3, 8,22,48,24,10	19, 6,37, 4,16,53,45,36
43	0	0	3, 8,31,50,52,52,50, 8	19, 5,42, 5, 0,11,17,12,14,52,31,45,28, 7,30
26	7	0	3, 8,44,37,14,52,48	19, 4,24,23, 2,48,45
9	14	0	3, 8,57,24,28,48	19, 3, 7, 6,20,14,48,53,2C
0	25	4	3, 9,10,12,34,41,57,11,15	19, 1,49,44,52, 8,10,37,47,20,14,15,58, 1,28,53,20
5	0	24	3, 9,15,51,40,40, 8,22,18,53,20	19, 1,15,39, 5,11,36,10,29,16, 6,31,40,48
16	0	4	3, 9,37,46,40	18,59, 3,45
1	8	5	3, 9,50,37,30	18,57,46,40
20	0	15	3,10,31,11, 3,22,28, 8,53,20	18,53,44,26,52,48
1	6	14	3,10,44, 5,30,28, 7,30	18,52,27,43,29,16,48
41	5	0	3,10,53,14,46, 2,29,45,36	18,51,3?,24,56,28,55,30,36,54,50,37,30
24	12	0	3,11, 6,10,42,48,57,36	18,50,16,50,25
7	19	0	3,11,19, 7,32, 9,36	18,49, 0,21, 4,26,29, 1,33,49,37,46,40
0	31	5	3,11,32, 5,14, 7,58,39, 8,26,15	18,47,43,56,54,27,20, 7,41,34, 3,43,10,38,30, 0,49,22,57,46,40
1	4	23	3,11,37,48,34,25,38,28,35,37,30	18,47,10,16,22,54,25,21,28, 9,59, 2,24
35	0	6	3,11,47, 0,24,41,28,53,20	18,46,16,13, 3,32,41,43, 7,3C
6	1	0	3,12	18,45
1	14	6	3,12,13, 0,28, 7,30	18,43,43,52, 5,55,33,20
8	0	10	3,12,54, 4,26,40	18,39,44,38,24
56	3	0	3,13, 3,19,56,32,59, 0,59,31,12	18,38,5C,56,56,54, 8,49,55,46,36,38,18,33,34,27,11,15
1	12	15	3,13, 7, 8,34,35,58,35,37,30	18,38,28,51,50,24
39	10	0	3,13,16,24,42, 7, 1,37,55,12	18,37,35,13,31,20,25,11,43, 7,30
22	17	0	3,13,29,30,20,51, 4,19,12	18,36,19,35,43,12,35,33,2C
5	24	0	3,13,42,36,52,48,43,12	18,35, 4, 3, 2, 9,51,37,50,26,47,40,54,19,15,33,20
12	0	21	3,13,48,24, 7, 5, 6,10,22,13,20	18,34,3C,45,12, 6,1C,28,59,31,12
23	0	1	3,14,10,50,40	18,32,21,56,36, 5,37,30
4	6	0	3,14,24	18,31, 6,40
1	20	7	3,14,37,10,13,28,35,37,30	18,29,51,28,29,33,23,17,31,51, 6,40
27	0	12	3,15, 5,31,57,41,43,42,13,20	18,27,1C, 7,30

```
 P  Q  R
 0  2  7   3,15,18,45
37 15  0   3,15,41,22, 0,38,36,54, 8,38,24
20 22  0   3,15,54,37,28,36,42,37,26,24
 3 29  0   3,16, 7,53,50,28,19,44,24
 0  0 16   3,16,13,45,25,10,25
42  0  3   3,16,23,10,30, 5, 2,13,20
19  4  0   3,16,36,28,48
 2 11  0   3,16,49,48
 1 26  8   3,17, 3, 8, 6, 8,42, 4,13, 7,3C
15  C  7   3,17,31,51, 6,40
 0  8  8   3,17,45,14, 3,45
18 27  0   3,18,21,33,26,43,10, 9,24,28,48
19  0 18   3,18,27,29, 1, 0,54,19,15,33,20
 1 34  0   3,18,34,59,45,51,10,59,12,18
 0  6 17   3,18,40,55,44,14,17,48,45
34  2  0   3,18,50,27,52,57,36
17  9  0   3,19, 3,56, 9,36
 C 16  0   3,19,17,25,21
34  0  9   3,19,46,27,55,43,12,35,33,20
 3  0  2   3,20
 0 14  9   3,20,13,32,59,17,48,45
 7  0 13   3,20,56,19,37,46,40
49  0  0   3,21, 5,58,16,24,21,28,32
 0 12 18   3,21, 9,56,26, 2,28,32, 6,33,45
32  7  0   3,21,19,35,43,52,19,12
15 14  0   3,21,33,14, 6,43,12
 0 22  1   3,21,46,53,25, 0,45
22  0  4   3,22,16,17,46,40
 1  5  2   3,22,30
 0 20 10   3,22,43,43, 9, 2,17, 6,33,45
26  0 15   3,23,13,15,47,35,58, 1,28,53,20
 1  3 11   3,23,27, 1,52,30
47  5  0   3,23,36,47,45, 6,39,44,38,24
30 12  0   3,23,50,35,25,40,13,26,24
13 19  0   3,24, 4,24, 2,18,14,24
 0 28  2   3,24,18,13,35, 4,30,33,45
 1  1 20   3,24,24,19,48,43,21, 2,30
41  0  6   3,24,34, 8,26,20,14,48,53,20
12  1  0   3,24,48
 1 11  3   3,25, 1,52,30
14  C 10   3,25,45,40,44,26,40
 1  9 12   3,25,59,37, 8,54,22,30
45 10  0   3,26, 9,30,20,55,29,44,26,52,48
28 17  0   3,26,23,28,22,14,28,36,28,48
11 24  0   3,26,37,27,20,19,58, 4,48
 0 34  3   3,26,51,27,15,15,48,56,40,18,45
 1  7 21   3,26,57,38, 3,34,53,33,16,52,30
29  0  1   3,27, 7,34, 2,40
10  6  0   3,27,21,36
 1 17  4   3,27,35,38,54,22,30
33  0 12   3,28, 5,54, 5,32,30,37, 2,13,20
 2  0  5   3,28,20
 1 15 13   3,28,34, 6,51,46, 3,16,52,30
26 22  0   3,28,58,15,58,31, 9,27,56, 9,36
 9 29  0   3,29,12,25,25,50,13, 3,21,36
 6  0 16   3,29,18,40,26,51, 6,40
48  0  3   3,29,28,43,12, 5,22,22,13,20
25  4  0   3,29,42,54,43,12
 8 11  0   3,29,57, 7,12
 1 23  5   3,30,11,20,38,33,16,52,30
21  0  7   3,30,41,58,31, 6,40
 0  5  5   3,30,56,15
 7 34  0   3,31,49,19,44,54,35,43, 9, 7,12
 0  3 14   3,31,55,39,27,11,15
40  0  0   3,32, 5,49,44,29,26,24
23  9  0   3,32,20,11,54,14,24
 6 16  0   3,32,34,35, 2,24
 1 29  6   3,32,48,59, 9, 2,11,50, 9,22,30
 0  1 23   3,32,55,20,38,15, 9,25, 6,15
40  0  9   3,33, 5,33,47,26, 5,25,55,33,20
 9  0  2   3,33,20
 0 11  6   3,33,34,27,11,15
13  0 13   3,34,20, 4,56,17,46,40
55  0  0   3,34,30,22, 9,29,58,54,26, 8
 0  9 15   3,34,34,36,11,46,38,26,15
38  7  0   3,34,44,54, 6,47,48,28,48
21 14  0   3,34,59,27, 3,10, 4,48
 4 21  0   3,35,14, 0,58,40,48
28  0  4   3,35,45,22,57,46,40
 3  3  0   3,36
 0 17  7   3,36,14,38, 1,38,26,15
 1  0  8   3,37, 0,50
53  5  0   3,37,11,14,56, 7, 6,23,36,57,36
 0 15 16   3,37,15,32, 8,55,28,25, 4,41,15
36 12  0   3,37,25,57,47,22,54,20, 9,36
19 19  0   3,37,40,41,38,27,27,21,36
 2 26  0   3,37,55,26,29,24,48,36
 5  0 19   3,38, 1,57, 7,58,14,26,40
47  0  6   3,38,12,25, 0, 5,35,48, 8,53,20
18  1  0   3,38,27,12
 1  8  0   3,38,42
 0 23  8   3,38,56,49, 0, 9,40, 4,41,15
20  0 10   3,39,28,43,27,24,26,40
 1  6  9   3,39,43,35,37,30
34 17  0   3,40, 9, 2,15,43,26,30,54,43,12
17 24  0   3,40,23,57, 9,41,17,57, 7,12
 0 31  0   3,40,38,53, 4,16,52,12,27
 1  4 18   3,40,45,28,35,49,13, 7,30
35  0  1   3,40,56, 4,18,50,40
16  6  0   3,41,11, 2,24
```

```
18,25,55,12
18,23,47,22,59, 6, 5,37,3C
18,22,32,41,12,18,21,46,59,45,11, 6,40
18,21,18, 4,28,48,15,26,15,44,58,56,41,47,54,37,21,58,31, 6,40
18,20,45,11,18,37,12,34,33,36
18,19,52,24, 0,10,50, 6,57,28,49,41,15
18,18,37,58, 7,3C
18,17,23,37,17, 2,13,20
18,16, 5,21,28,27, 3, 0,16,38,37,41,43,42,13,20
18,13,3C
18,12,16
18, 8,55,59,12,53,41,30,51,36,28,45, 6,10,22,13,20
18, 8,23,28,12,17,16,48
18, 7,42,17,45,29, 8,34,49,37,45,37,28,41,23,34,40,57,47,45,50,37, 2,13,20
18, 7, 9,48,56,54,31,40,48
18, 6,17,40,44,37,22, 5,23,26,15
18, 5, 4,10
18, 3,50,44,13,51,49,27,54, 4,26,40
18, 1,13,10, 8,12,11,15
18
17,58,46,54,48,53,20
17,54,57,15,15,5C,24
17,54, 5,42,11,25,34,52,43,56,44,46,22,37, 1,52,30
17,53,44,30,33,59, 2,24
17,52,53, 0,58,53,12,11,15
17,51,4C,24,41,28,53,20
17,50,27,53,18,52,39,57,55,37,43,22,28, 8,53,20
17,47,52,15,56,15
17,46,4C
17,45,27,48,57,10,27, 9,37,46,40
17,42,52,55,12
17,41,40,59,31,12
17,40,5C, 4,37,57, 7, 2,27, 6,24,57,39,22,30
17,39,38,17,15,56,15
17,38,26,34,45,24,49,42,42,57,46,40
17,37,14,57, 6, 3, 7,37,12,43,10,59,13,43,35,38,16,17,46,40
17,36,43,22,51,28,31,16,22,29,21,36
17,35,52,42,14,34,24, 6,40,46,52,30
17,34,41,15
17,33,29,52,35,33,2C
17,29,45,36
17,28,34,23,36
17,27,44,16,25,37,53,37,14,10,46,52,30
17,26,33,22,14,15,33,20
17,25,22,32,5C,46,44,39,13,32,37,12, 5,r5,33,20
17,24,11,48,14,51,58,38,14, 2,38,59,58,44,32,14, 5,43,29, 3,12,35,33,20
17,23,40,37,23,25,56,48,46, 4,48
17,22,5C,34,18,50,16,24,22,30
17,21,4C
17,20,29,30,27,42,33, 5,11, 6,40
17,17,56,14,31,52,30
17,16,48
17,15,37,50,13,20
17,13,38, 8,37,47,12,55,18,31, 6,40
17,12,28,11,42, 0,14,28,22,15,55,15,39,11, 9,57,31,51, 6,40
17,11,57,21,51,12,23, 2,24
17,11, 7,52,3C,10, 9,29, 1,23,16,34,55,18,45
17, 9,56, 5,44,31,52,30
17, 8,48,23,42,13,20
17, 7,38,46,22,55,21,34, C,36,12,52,50,22,13,20
17, 5, 9,22,30
17, 4
16,59,43,24, 8,53,34,17,39, 1,39, 1,23, 8,48,21,15,54,11, 1,43,42,13,20
16,59,12,57, 8,21, 7,12
16,58,24, 4,26,50, 1,57,33,13,21,33,45
16,57,15, 9,22,30
16,56, 6,18,57,59,5C, 7,24,26,40
16,54,57,33,13, 0,36, 6,55,24,39,20,51,34,39, 0,44,26,40
16,54,27,14,44,36,58,49,19,20,59, 8, 9,36
16,53,38,35,45,11,25,32,48,45
16,52,3C
16,51,21,28,53,2C
16,47,46,10,33,36
16,46,57,50,48,12,43,56,56,11,56,58,28,42,13, 0,28, 7,30
16,46,37,58,39,21,36
16,45,49,42,10,12,22,40,32,48,45
16,44,41,38, 8,53,20
16,43,33,38,43,56,52,28, 3,24, 6,54,48,53,20
16,41, 7,44,56,29, 3,45
16,40
16,38,52,19,38,36, 2,57,46,40
16,35,19,40,48
16,34,31,56,50,34,47,51, 2,54,45,54, 3, 9,50,37,30
16,34,12,19,24,48
16,33,24,38,41,11,29, 3,45
16,32,17,25, 5, 4,31,36,17,46,40
16,31,10,16, 1,55,25,53,38,10,29, 3, 1,37, 7, 9,37,46,40
16,30,4C,40,10,45,29,19, 6,14,24
16,29,53, 9,36, 9,45, 6,15,43,56,43, 7,30
16,28,46,10,18,45
16,27,35,15,33,20
16,26,32,25,19,36,20,42,14,58,45,55,33,20
16,24, 9
16,23, 2,24
16,21, 8,47, 5,52, 5
16,20, 2,23,17,36,19,21,46,26,49,52,35,33,20
16,18,56, 3,58,56,13,43,20,39,59, 3,43,49,15,13,12,52, 0,59,15,33,20
16,18,2C,50, 3,13, 4,30,43,12
16,17,35,54,40, 9,37,52,51, 5,37,30
16,16,33,45
```

P	Q	R		
1	14	1	3,41,26, 1,30	16,15,27,39,48,28,38,31, 6,40
8	0	5	3,42,13,20	16,12
1	12	10	3,42,28,23,19,13, 7,30	16,10,54,13,20
15	29	0	3,43, 9,15, 7,33,33,55,35, 2,24	16, 7,56,25,58, 7,43,34, 5,52,25,33,25,29,13, 5,11, 6,40
12	0	16	3,43,15,55, 8,38,31, 6,40	16, 7,27,31,44,15,21,36
0	37	1	3,43,24,22,14, 5, 4,51,36,20,15	16, 6,50,55,47, 5,54,17,37,26,53,53,18,50, 7,37,29,44,42,27,24,59,35,18,31, 6,40
54	0	3	3,43,26,38, 4,53,43,51,42,13,20	16, 6,41, 7,58,17, 1,23,27,33, 4,17,44,21,19,41,15
1	10	19	3,43,31, 2,42,16, 5, 2,20,37,30	16, 6,22, 3,30,35, 8, 9,36
31	4	0	3,43,41,46,22, 4,48	16, 5,35,42,52,59,52,58, 7,30
14	11	0	3,43,56,55,40,48	16, 4,30,22,13,20
1	20	2	3,44,12, 6, 1, 7,30	16, 3,25, 5,58,59,23,58, 8, 3,57, 2,13,20
27	0	7	3,44,44,46,25,11, 6,40	16, 1, 5, 2,20,37,30
0	2	2	3,45	16
1	18	11	3,45,15,14,36,42,32,20,37,30	15,58,55, 2, 3,27,24,26,40
0	0	11	3,46, 3,22, 5	15,55,30,53,34, 4,48
46	2	0	3,46,14,13, 3,27,24, 9,36	15,54,45, 4,10, 9,24,20,12,23,46,27,53,26,15
29	9	0	3,46,29,32,41,51,21,36	15,53,40,27,32,20,37,30
12	16	0	3,46,44,53,22,33,36	15,52,35,55,16,52,20,44,26,40
1	26	3	3,47, 0,15, 5,38,20,37,30	15,51,31,27,23,26,48,51,29,26,51,53,18,21,14, 4,26,40
4	0	22	3,47, 7, 2, 0,48,10, 2,46,40	15,51, 3, 2,34,19,40, 8,44,23,25,26,24
15	0	2	3,47,33,20	15,49,13, 7,30
0	8	3	3,47,48,45	15,48, 8,53,20
19	0	13	3,48,37,25,16, 2,57,46,40	15,44,47, 2,24
61	0	0	3,48,48,23,38, 7,58,50, 3,52,32	15,44, 1,43,52,41,56,12, 7,41,12, 9,49,24,34,41,41,22, 1,52,30
0	6	12	3,48,52,54,36,33,45	15,43,43, 6,14,24
44	7	0	3,49, 3,53,43,14,59,42,43,12	15,42,57,50,47, 4, 6,15,30,45,42,11,15
27	14	0	3,49,19,24,51,22,45, 7,12	15,41,54, 2, 0,50
10	21	0	3,49,34,57, 2,35,31,12	15,40,50,17,33,42, 4,11,18,11,21,28,53,20
1	32	4	3,49,50,30,16,57,34,22,58, 7,30	15,39,46,37,25,22,46,46,24,38,23, 5,58,52, 5, 0,41, 9, 8, 8,53,20
0	4	21	3,49,57,22,17,18,46,10,18,45	15,39,18,33,39, 5,21, 7,53,28,19,12
34	0	4	3,50, 8,24,29,37,46,40	15,38,33,30,52,57,14,45,56,15
9	3	0	3,50,24	15,37,30
0	14	4	3,50,39,36,33,45	15,36,26,33,24,56,17,46,40
7	0	8	3,51,28,53,20	15,33, 7,12
0	12	13	3,51,44,34,17,31,10,18,45	15,32, 4, 3,12
42	12	0	3,51,55,41,38,32,25,57,30,14,24	15,31,19,21,16, 7, 0,59,45,56,15
25	19	0	3,52,11,24,25, 1,17,11, 2,24	15,30,16,19,46, 0,29,37,46,40
8	26	0	3,52,27, 8,15,22,27,50,24	15,29,13,22,31,48,13, 1,32, 2,19,44, 5,16, 2,57,46,40
11	0	19	3,52,34, 4,56,30, 7,24,26,40	15,28,45,37,40, 5, 8,44, 9,36
24	1	0	3,53, 1, 0,48	15,26,58,17,10, 4,41,15
7	8	0	3,53,16,48	15,25,55,33,20
0	20	5	3,53,32,36,16,10,18,45	15,24,52,53,44,37,49,24,36,32,35,33,20
26	0	10	3,54, 6,38,21,14, 4,26,40	15,22,38,26,15
1	3	6	3,54,22,30	15,21,36
0	18	14	3,54,38,22,43,14,18,41,29, 3,45	15,20,33,37,58,31, 6,40
23	24	0	3,55, 5,32,58,20, 3, 8,55,40,48	15,18,47,14,20,15,18, 9, 9,47,39,15,33,20
6	31	0	3,55,21,28,36,33,59,41,16,48	15,17,45, 3,44, 0,12,51,53, 7,29, 7,14,49,55,31, 8,18,45,55,33,20
1	1	15	3,55,28,30,30,12,30	15,17,17,39,25,31, 0,28,48
41	0	1	3,55,39,48,36, 6, 2,40	15,16,33,40, 0, 9, 1,45,47,54, 1,24,22,30
22	6	0	3,55,55,46,33,36	15,15,31,38,26,15
5	13	0	3,56,11,45,36	15,14,29,41, 4,11,51, 6,40
0	26	6	3,56,27,45,43,22,26,29, 3,45	15,13,27,47,53,42,32,30,13,52,11,24,46,25,11, 6,40
3	0	25	3,56,34,49,35,50,10,27,53,36,40	15,13, 0,31,16, 9,16,56,23,24,53,13,20,38,24
14	0	5	3,57, 2,13,20	15,11,15
1	9	7	3,57,18,16,52,30	15,10,13,20
18	0	16	3,58, 8,58,49,13, 5,11, 6,40	15, 6,55,33,30,14,24
4	36	0	3,58,17,59,43, 1,25,11, 2,45,36	15, 6,25,14,47,54,17, 9, 1,21,28, 1,13,54,29,38,54, 8, 9,48,12,10,51,51, 6,40
1	7	16	3,58,25, 6,53, 5, 9,22,30	15, 5,58,10,47,25,26,24
37	4	0	3,58,36,33,27,33, 7,12	15, 5,14,43,57,11, 8,24,29,31,52,30
20	11	0	3,58,52,43,23,31,12	15, 4,13,28,20
3	18	0	3,59, 8,54,25,12	15, 3,12,16,51,33,11,13,15, 3,42,13,20
33	0	7	3,59,43,45,30,51,51, 6,40	15, 1, 0,58,26,50, 9,22,30
2	0	0	4	15
1	15	8	4, 0,16,15,35, 9,22,30	14,58,59, 5,40,44,26,40
6	0	11	4, 1, 7,35,33,20	14,55,47,42,43,12
52	2	0	4, 1,19, 9,55,41,13,46,14,24	14,55, 4,45, 9,31,19, 3,56,37,17,18,38,50,51,33,45
1	13	17	4, 1,23,55,43,14,58,14,31,52,30	14,54,47, 5,28,19,12
35	9	0	4, 1,35,30,52,38,47, 2,24	14,54, 4,10,49, 4,20, 9,22,30
18	16	0	4, 1,51,52,56, 3,50,24	14,53, 3,40,34,34, 4,26,40
1	23	0	4, 2, 8,16, 6, 0,54	14,52, 3,14,25,43,53,18,16,21,26, 8,43,27,24,26,40
10	0	22	4, 2,15,30, 8,51,22,42,57,46,40	14,51,36,36, 9,40,56,23,11,36,57,36
21	0	2	4, 2,43,33,20	14,49,53,33,16,52,30
0	5	0	4, 3	14,48,53,20
1	21	9	4, 3,16,27,46,50,44,31,52,30	14,47,53,10,47,38,42,38, 1,28,53,20
25	0	13	4, 3,51,54,57, 7, 9,37,46,40	14,45,44, 6
0	3	9	4, 4, 8,26,15	14,44,44, 9,36
50	7	0	4, 4,20, 9,18, 7,59,41,34, 4,48	14,44, 1,43,51,37,35,52, 2,35,20,48, 2,48,45
33	14	0	4, 4,36,42,30,48,16, 7,40,48	14,43, 1,54,23,16,52,30
16	21	0	4, 4,53,16,50,45,53,16,48	14,42, 2, 8,57,50,41,25,35,48, 8,53,20
1	29	1	4, 5, 9,52,18, 5,24,40,30	14,41, 2,27,35, 2,36,21, 0,35,59, 9,21,76,19,41,53,34,48,53,20
0	1	18	4, 5,17,11,46,28, 1,15	14,40,36, 9, 2,53,46, 3,38,52,48
40	0	4	4, 5,28,58, 7,36,17,46,40	14,39,53,55,12, 8,40, 5,33,59, 3,45
15	3	0	4, 5,45,36	14,38,54,22,30
0	11	1	4, 6, 2,15	14,37,54,53,49,37,46,40
13	0	8	4, 6,54,48,53,20	14,34,48
0	9	10	4, 7,11,32,34,41,15	14,33,48,48
31	19	0	4, 7,40,10, 2,41,22,19,46,23,36	14,32, 7,48,31,52,57,46,40
14	26	0	4, 7,56,56,48,23,57,41,45,36	14,31, 8,47,22,18,57,12,41,17,11, 0, 4,56,17,46,40
17	0	19	4, 8, 4,21,16,16, 7,54, 4,26,40	14,30,42,46,33,49,49,26,24
1	35	2	4, 8,13,44,42,18,58,44, 0,22,30	14,30, 9,50,12,23,18,51,51,42,12,29,58,57, 6,51,44,46,14,12,40,29,37,46,4
0	7	19	4, 8,21, 9,40,17,52,15,56,15	14,29,43,51, 9,31,37,20,38,24
30	1	0	4, 8,33, 4,51,12	14,29, 2, 8,35,41,53,40,18,45
13	8	0	4, 8,49,55,12	14,28, 3,20
0	17	2	4, 9, 6,46,41,15	14,27, 4,35,23, 5,27,34,19,15,33,20
32	0	10	4, 9,43, 4,54,39, 0,44,26,40	14,24,58,32, 6,33,45
1	0	3	4,10	14,24
0	15	11	4,10,16,56,14, 7,15,56,15	14,23, 1,31,51, 6,40
12	31	0	4,11, 2,54,31, 0,15,40, 1,55,12	14,20,23,29,45, 0,12, 3,38,33,16, 3, 2,39,18,17,56,32,35,33,20
5	0	14	4,11,10,24,32,13,20	14,19,57,48,12,40,19,12
47	0	1	4,11,22,27,50,30,26,50,40	14,19,16,23,45, 8,27,54,11, 9,23,49, 6, 5,37,30

```
P   Q   R
28  6   0   4,11,39,29,39,50,24                 14,18,18,24,47, 6,33,45
11 13   0   4,11,56,32,38,24                    14,17,20,19,45,11, 6,40
 0 23   3   4,12,13,36,46,15,56,15              14,16,22,18,39, 6, 7,58,20,30,10,41,58,31, 6,40
20  0   5   4,12,50,22,13,20                    14,14,17,48,45
 1  6   4   4,13, 7,30                          14,13,2C
 0 21  12   4,13,24,38,56,17,51,23,12,11,15     14,12,22,15, 9,44,21,43,42,13,20
24  0  16   4,14, 1,34,44,29,57,31,51, 6,40     14,10,18,20, 9,36
 1  4  13   4,14,18,47,20,37,30                 14, 9,20,47,36,57,36
43  4   0   4,14,30,59,41,23,19,40,48           14, 8,4C, 3,42,21,41,37,57,41, 7,58, 7,30
26 11   0   4,14,48,14,17, 5,16,48              14, 7,42,37,48,45
 9 18   0   4,15, 5,30, 2,52,48                 14, 6,45,15,48,19,51,46,10,22,13,20
 0 29   4   4,15,22,46,58,50,38,12,11,15        14, 5,47,57,40,50,30, 5,46,10,32,47,22,58,52,30,37, 2,13,20
 1  2  22   4,15,30,24,45,54,11,18, 7,30        14, 5,22,42,17,10,49, 1, 6, 7,29,16,48
39  0   7   4,15,42,40,32,55,18,31, 6,40        14, 4,42, 9,47,39,31,17,20,37,30
 8  0   0   4,16                                14, 3,45
 1 12   5   4,16,17,20,37,30                    14, 2,47,54, 4,26,40
12  0  11   4,17,12, 5,55,33,20                 13,59,48,28,48
58  2   0   4,17,24,26,35,23,58,41,19,21,36     13,59, 8,12,20,10,36,37,26,49,57,28,43,55,10,50,23,26,15
 1 10  14   4,17,29,31,26, 7,58, 7,30           13,58,51,38,52,48
41  9   0   4,17,41,52,56, 9,22,10,33,36        13,58,11,25, 8,30,18,53,47,20,37,30
24 16   0   4,17,59,20,27,48, 5,45,36           13,57,14,41,47,24,26,40
 7 23   0   4,18,16,49,10,24,57,36              13,56,18, 2,16,37,23,43,22,50, 5,45,40,44,26,40
27  0   2   4,18,54,27,33,20                    13,54,16,27,27, 4,13, 7,30
 6  5   0   4,19,12                             13,53,20
 1 18   6   4,19,29,33,37,58, 7,30              13,52,23,36,22,10, 2,28, 8,53,20
31  0  13   4,20, 7,22,36,55,38,16,17,46,40     13,50,22,35,37,30
 0  0   6   4,20,25                             13,49,26,24
 1 16  15   4,20,42,38,34,42,34, 6, 5,37,30     13,48,30,16,1C,40
39 14   0   4,20,55, 9,20,51,29,12,11,31,12     13,46,54,30,54,13,46,20,14,48,53,20
22 21   0   4,21,12,49,58, 8,56,49,55,12        13,45,58,33,21,36,11,34,41,48,44,12,31,20,55,58, 1,28,53,20
 5 28   0   4,21,30,31,47,17,46,19,12           13,45,33,53,28,57,54,25,55,12
 4  0  17   4,21,38,20,33,33,53,20              13,44,54,18, 0, 8, 7,35,13, 6,37,15,56,15
46  0   4   4,21,50,54, 0, 6,42,57,46,40        13,43,58,28,35,37,3C
21  3   0   4,22, 8,38,24                       13,43, 2,42,57,46,40
 4 10   0   4,22,26,24                          13,42, 7, 1, 6,20,17,15,12,28,58,16,17,46,40
 1 24   7   4,22,44,10,48,11,36, 5,37,30        13,40, 7,30
19  0   8   4,23,22,28, 8,53,20                 13,39,12
 0  6   7   4,23,40,18,45                       13,36,41,59,24,40,16, 8, 8,42,21,33,49,37,46,40
20 26   0   4,24,28,44,35,37,33,32,32,38,24     13,35,46,43,19, 6,51,26, 7,13,19,13, 6,31, 2,41, 0,43,20,49,22,57,46,40
 3 33   0   4,24,46,39,41, 8,14,38,56,24        13,35,22,21,42,40,53,45,36
 0  4  16   4,26,54,34,18,59, 3,45              13,34,43,15,33,28, 1,34, 2,34,41,15
36  1   0   4,25, 7,17,10,36,48                 13,33,48, 7,30
19  8   0   4,25,25,14,52,48                    13,32,53, 3,10,23,52, 5,55,33,20
 2 15   0   4,25,43,13,48                       13,31,33,47,47,41,35, 3,27,28,47,18,31,40,48
 0  2  25   4,26, 9,10,47,48,56,46,22,48,45     13,30,54,52,36, 9, 8,26,15
38  0  10   4,26,21,57,14,17,36,47,24,26,40     13,30
 7  0   3   4,26,40                             13,29, 5,11, 6,40
 0 12   8   4,26,58, 3,59, 3,45                 13,26,12,56,26,52,48
11  0  14   4,27,55, 6,10,22,13,20              13,25,42,26,29,14,55,14,41,12,24,54,25,41,46,21,14,47,15,22,50,49,39,25,25,55,33,20
 1 38   0   4,28, 5,14,40,56, 5,49,55,36,18     13,25,34,16,38,34,11, 9,32,57,33,34,46,57,46,24,22,30
53  0   1   4,28, 7,57,41,52,28,38, 2,40        13,25,18,22,55,29,16,48
 0 10  17   4,28,13,15,14,43,18, 2,48,45        13,24,35,45,44, 9,54, 8,26,15
34  6   0   4,28,26, 7,38,29,45,36              13,23,45,18,31, 6,4C
17 13   0   4,28,44,18,48,57,36                 13,22,5C,54,59, 9,29,58,26,43,17,31,51, 6,40
 0 20   0   4,29, 2,31,13,21                    13,2C,54,11,57,11,15
26  0   5   4,29,41,43,42,13,20                 13,20
 1  3   1   4,30                                13,19, 5,51,42,52,50,22,13,20
 0 18   9   4,30,18,17,32, 3, 2,48,45           13,16,15,44,38,24
 1  1  10   4,31,16, 2,30                       13,15,37,23,28,27,50,16,50,19,48,43,14,31,52,30
49  4   0   4,31,29, 3,40, 8,52,59,31,12        13,14,43,42,56,57,11,15
32 11   0   4,31,47,27,14,13,37,55,12           13,14,49,56, 4, 3,37,17, 2,13,20
15 18   0   4,32, 5,52, 3, 4,19,12              13,12,5C,12,49,32,20,42,54,32,23,14,25,17,41,43,42,13,20
 0 26   1   4,32,24,18, 6,46, 0,45              13,12,32,32, 8,36,23,27,16,59,31,12
 3  0  20   4,32,32,26,24,57,48, 3,20           13,11,54,31,40,55,48, 5, 0,35, 9,22,30
45  0   7   4,32,45,31,15, 6,59,45,11, 6,40     13,11, 0,56,15
14  0   0   4,33, 4                             13,10, 7,24,26,40
 1  9   2   4,33,22,30                          13, 9,13,56,15,41, 4,33,47,59, 0,44,26,40
 0 24  10   4,33,41, 1,15,12, 5, 5,51,33,45     13, 7,19,12
18  0  11   4,34,20,54,19,15,33,20              13, 6,25,55,12
 1  7  11   4,34,39,29,31,52,30                 13, 5,48,12,19,13,25,12,55,38, 5, 9,22,30
47  9   0   4,34,52,40,27,53,59,39,15,50,24     13, 4,55, 1,40,41,40
30 16   0   4,35,11,17,49,39,18, 8,38,24        13, 4, 1,54,38, 5, 3,29,25, 9,27,54, 4,26,40
13 23   0   4,35,29,56,27, 6,37,26,24           13, 3, 8,51,11, 8,58,58,4C,31,59,14,59, 3,24,10,34,17,36,47,24,26,40
 0 32   2   4,35,48,36,20,21, 5,15,33,45        13, 2,45,28, 2,34,27,36,34,33,36
 1  5  20   4,36,50,44,46,31,24,22,30           13, 2, 7,55,44, 7,42,18,16,52,30
33  0   2   4,36,10, 5,23,33,20                 13, 1,15
12  5   0   4,36,28,48                          13, 0,22, 7,50,46,54,48,53,20
 1 15   3   4,36,47,31,52,30                    12,57,36
 6  0   6   4,37,46,40                          12,56,43,22,40
 1 13  12   4,38, 5,29, 9, 1,24,22,30           12,55,12,36,28,20,24,41,28,53,20
28 21   0   4,38,37,41,18, 1,32,37,14,52,48     12,54,21, 8,46,30,10,51,16,41,56,26,44,23,22,28, 8,53,20
11 28   0   4,38,56,33,54,26,57,24,28,48        12,53,20,54,22,37,37, 6,46, 2,27,26,11,79, 3,45
10  0  17   4,39, 4,53,55,48, 8,53,20           12,52,28,34,18,23,54,22,3C
52  0   4   4,39,18,17,36, 7, 9,49,37,46,40     12,51,36,17,46,40
27  3   0   4,39,37,12,57,36                    12,50,44, 4,17,31,10,30,27, 9,37,46,40
10 10   0   4,39,56, 9,36                       12,48,52, 1,52,30
 1 21   4   4,40,15, 7,31,24,22,30              12,48
25  0   8   4,40,55,58, 1,28,53,20              12,47, 8, 1,38,45,55,33,2C
 0  3   4   4,41,15                             12,44,47,33,49, 6,40,10,43,14,16,14,16, 2,21,36,15,56,55,38,16,17,46,40
 1 19  13   4,41,34, 3,15,53,10,25,46,52,30     12,44,24,42,51,15,5C,24
 9 33   0   4,42,25,46,19,52,47,37,32, 9,36     12,43,48, 3,20,71,31,28, 9,55, 1,10,18,45
 0  1  13   4,42,34,12,36,15                    12,42,5C,22, 1,52,30
42  1   0   4,42,47,46,19,19,15,12              12,42, 4,44,13,29,52,35,33,20
25  8   0   4,43, 6,55,52,19,12                 12,41,13, 9,54,45,27, 5,11,33,29,30,38,40,59,15,33,20
 8 15   0   4,43,26, 6,43,12                    12,40,5C,26, 3,27,44, 6,59,30,44,21, 7,12
 1 27   5   4,43,45,18,52, 2,55,46,52,30        12,39,22,30
 2  0  23   4,43,53,47,31, 0,12,33,28,20        12,38,31, 6,40
13  0   3   4,44,26,40                          12,35,49,37,55,12
 0  9   5   4,44,45,56,15
17  0  14   4,45,46,46,35, 3,42,13,20
```

P	Q	R		
59	0	1	4,46, 0,29,32,39,58,32,34,50,40	12,35,13,23, 6, 9,32,57,42, 8,57,43,51,31,39,45,21, 5,37,30
0	7	14	4,46, 6, 8,15,42,11,15	12,34,56,28,59,31,12
40	6	0	4,46,19,52, 9, 3,44,38,24	12,34,22,16,37,39,17, 0,24,36,33,45
23	13	0	4,46,39,16, 4,13,26,24	12,33,31,13,26,40
6	20	0	4,46,58,41,18,14,24	12,32,4C,14, 2,57,39,21, 2,33, 5,11, 6,40
0	5	23	4,47,26,42,51,38,27,42,53,26,15	12,31,26,50,55,16,16,54,18,46,39,21,36
32	0	5	4,47,40,30,37, 2,13,20	12,30,5C,48,42,21,47,48,45
5	2	0	4,48	12,30
0	15	6	4,48,19,30,42,11,15	12,29, 9,14,43,57, 2,13,20
5	0	9	4,49,21, 6,40	12,26,29,45,36
55	4	0	4,49,34,59,54,49,28,31,29,16,48	12,25,53,57,37,56, 5,53,17,11, 4,25,32,22,22,58, 7,30
0	13	15	4,49,40,42,51,53,57,53,26,15	12,25,39,14,33,36
38	11	0	4,49,54,37, 3,10,32,26,52,48	12,25, 3,29, 0,53,36,47,48,45
21	18	0	4,50,14,15,31,16,36,28,48	12,24,13, 3,48,48,23,42,13,20
4	25	0	4,50,33,55,19,13, 4,48	12,23,22,42, 1,26,34,25,13,37,51,47,16,12,50,22,13,20
9	0	20	4,50,42,36,10,37,39,15,33,20	12,23, C,30, 8, 4, 6,59,19,40,48
20	0	0	4,51,16,16	12,21,34,37,44, 3,45
3	7	0	4,51,36	12,20,44,26,40
0	21	7	4,51,55,45,20,12,53,26,15	12,19,54,18,59,42,15,31,41,14, 4,26,40
24	0	11	4,52,38,17,56,32,35,33,2C	12,18, 6,45
1	4	8	4,52,58, 7,30	12,17,16,48
36	16	0	4,53,32, 3, 0,57,55,21,12,57,36	12,15,51,35,19,24, 3,45
19	23	0	4,53,51,56,12,55, 3,56, 9,36	12,15, 1,47,28,12,14,31,19,50, 7,24,26,40
2	30	0	4,54,11,50,45,42,29,36,36	12,14,12, 2,59,12,10,17,30,29,59,17,47,51,56,24,54,39, 0,44,26,40
1	2	17	4,54,20,38, 7,45,37,30	12,13,50, 7,32,24,48,23, 2,24
39	0	2	4,54,34,45,45, 7,33,20	12,13,14,56, 0, 7,13,24,38,19,13, 7,30
18	5	0	4,54,54,43,12	12,12,25,18,45
1	12	0	4,55,14,42	12,11,35,44,51,21,28,53,20
0	27	8	4,55,34,42, 9,13, 3, 6,19,41,15	12,10,46,14,18,58, 2, 0,11, 5,45, 7,49, 8, 8,53,20
1	0	26	4,55,43,31,59,47,43, 4,52, 0,5C	12,10,24,25, 0,55,25,33, 6,43,54,34,40,70,43,12
12	0	6	4,56,17,46,40	12, 9
1	10	9	4,56,37,51, 5,37,30	12, 8,10,40
17	28	0	4,57,32,20,10, 4,45,14, 6,43,12	12, 5,57,19,28,35,47,40,34,24,19,10, 4, 6,54,48,53,20
16	0	17	4,57,41,13,31,31,21,28,53,20	12, 5,35,38,48,11,31,12
0	35	0	4,57,52,29,38,46,46,28,48,27	12, 5, 8,11,50,19,25,43,13, 5,10,24,59, 7,35,43, 7,18,31,50,33,44,41,28,53,20
1	8	18	4,58, 1,23,36,21,26,43, 7,30	12, 4,46,32,37,56,21, 7,12
33	3	0	4,58,15,41,49,26,24	12, 4,11,47, 9,44,54,43,35,37,30
16	10	0	4,58,35,54,14,24	12, 3,22,46,40
1	18	1	4,58,56, 8, 1,30	12, 2,33,49,29,14,32,58,36, 2,57,46,40
31	0	8	4,59,39,41,53,34,48,53,20	12, 0,48,46,45,28, 7,30
0	0	1	5	12
1	16	10	5, 0,20,19,28,56,43, 7,30	11,59,11,16,32,35,33,20
4	0	12	5, 1,24,29,26,40	11,56,38,10,10,33,36
48	1	0	5, 1,38,57,24,36,32,12,48	11,56, 3,48, 7,37, 3,15, 9,17,49,50,55, 4,41,15
31	8	0	5, 1,59,23,35,48,28,48	11,55,15,20,39,15,28, 7,30
14	15	0	5, 2,19,51,10, 4,48	11,54,26,56,27,39,15,33,20
1	24	2	5, 2,40,20, 7,31, 7,30	11,53,38,35,32,35, 6,38,37, 5, 8,54,58,45,55,33,20
8	0	23	5, 2,49,22,41, 4,13,23,42,13,20	11,53,17,16,55,44,45, 6,33,17,34, 4,48
19	0	3	5, 3,24,26,40	11,51,54,50,37,30
0	6	2	5, 3,45	11,51, 6,40
1	22	11	5, 4, 5,34,43,33,25,39,50,37,30	11,50,18,32,38, 6,58, 6,25,11, 6,40
23	0	14	5, 4,49,53,41,23,57, 2,13,20	11,48,35,16,48
0	4	11	5, 5,10,32,48,45	11,47,47,19,40,48
46	6	0	5, 5,25,11,37,39,59,36,57,36	11,47,13,23, 5,18, 4,41,38, 4,16,38,26,15
29	13	0	5, 5,45,53, 8,30,20, 9,36	11,46,25,31,3C,37,30
12	20	0	5, 6, 6,36, 3,27,21,36	11,45,37,43,10,16,33, 8,28,38,31, 6,40
1	30	3	5, 6,27,20,22,36,45,50,37,30	11,44,49,58, 4, 2, 5, 4,48,28,47,19,29, 9, 3,45,30,51,51, 6,40
0	2	20	5, 6,36,29,43, 5, 1,33,45	11,44,2C,55,14,19, 0,50,55, 6,14,24
38	0	5	5, 6,51,12,39,30,22,13,20	11,43,55, 8, 9,42,56, 4,27,11,15
11	2	0	5, 7,12	11,43, 7,30
0	12	3	5, 7,32,48,45	11,42,19,55, 3,42,13,20
11	0	9	5, 8,38,31, 6,40	11,39,50,24
0	10	12	5, 8,59,25,43,21,33,45	11,39, 3, 2,24
44	11	0	5, 9,14,15,31,23,14,36,40,19,12	11,38,29,30,57, 5,15,44,49,27,11,15
27	18	0	5, 9,35,12,33,21,42,54,43,12	11,37,42,14,49,30,22,13,20
10	25	0	5, 9,56,11, 0,29,57, 7,12	11,36,55, 1,53,51, 9,46, 9, 1,44,48, 3,57, 2,13,20
15	0	20	5,10, 5,26,35,20, 9,52,35,33,2C	11,36,34,13,15, 3,51,33, 7,12
0	8	21	5,10,26,27, 5,22,20,19,55,18,45	11,35,47, 4,55,37,17,52,30,43,12
26	0	0	5,10,41,21, 4	11,35,13,42,52,33,30,56,15
9	7	0	5,11, 2,24	11,34,26,40
0	18	4	5,11,23,28,21,33,45	11,33,39,40,18,28,22, 3,27,24,26,40
30	0	11	5,12, 8,51, 8,18,45,55,33,20	11,31,58,49,41,15
1	1	5	5,12,30	11,31,12
0	16	13	5,12,51,10,17,39, 4,55,18,45	11,30,25,13,28,53,20
25	23	0	5,13,27,23,57,46,44,11,54,14,24	11,29, 5,25,45,11,28,36,52,20,44,26,40
8	30	0	5,13,48,38, 8,45,19,35, 2,24	11,28,18,47,48, 0, 9,38,54,50,36,50,26, 7,26,38,21,14, 4,26,40
3	0	15	5,13,58, 0,40,16,40	11,27,58,14,34, 8,15,21,36
45	0	2	5,14,13, 4,48, 8, 3,33,20	11,27,25,15, 0, 6,46,19,20,55,31, 3,16,52,30
24	5	0	5,14,34,22, 4,48	11,26,38,43,49,41,15
7	12	0	5,14,55,40,48	11,25,52,15,48, 8,53,20
0	24	5	5,15,17, 0,57,49,55,18,45	11,25, 5,50,55,16,54,22,40,24, 8,33,34,48,53,20
18	0	6	5,16, 2,57,46,40	11,23,2C,15
1	7	6	5,16,24,22,30	11,22,40
22	0	17	5,17,31,58,25,37,26,54,48,53,2C	11,20,14,40, 7,40,48
6	35	0	5,17,43,59,37,21,53,34,43,40,48	11,19,48,56, 5,55,42,51,46, 1, 6, 0,55,25,52,14,10,36, 7,21, 9, 8, 8,53,20
1	5	15	5,17,53,29,10,46,52,30	11,19,28,38, 5,34, 4,48
39	3	0	5,18, 8,44,36,44, 9,36	11,18,56, 2,57,53,21,18,22, 8,54,22,30
22	10	0	5,18,30,17,51,21,36	11,18,10, 6,15
5	17	0	5,18,51,52,33,36	11,17,24,12,38,39,53,24,56,17,46,40
0	30	6	5,19,13,28,43,33,17,45,14, 3,45	11,16,38,22, 8,40,24, 4,36,56,26,13,54,23, 6, 0,29,37,46,40
1	3	24	5,19,23, 0,57,22,44, 7,39,22,30	11,16,18, 9,49,44,39,12,52,53,59,25,26,24
37	0	8	5,19,38,20,41, 9, 8, 8,53,20	11,15,45,43,50, 7,37, 1,52,30
6	0	1	5,20	11,15
0	13	7	5,20,21,40,46,52,30	11,14,14,19,15,33,20
10	0	12	5,21,30, 7,24,26,40	11,13,5C,47, 2,24
54	1	0	5,21,45,33,14,14,58,21,39,12	11,11,18,33,52, 8,29,17,57,27,57,58,59, 8, 8,40,18,45
1	11	16	5,21,51,54,17,39,57,39,22,30	11,11, 5,19, 6,14,24
37	8	0	5,22, 7,21,10,11,42,43,12	11,10,33, 8, 6,48,15, 7, 1,52,30
20	15	0	5,22,29,10,34,45, 7,12	11, 9,47,45,25,55,33,20
3	22	0	5,22,51, 1,28, 1,12	11, 9, 7,25,49,17,54,58,42,16, 4,36,32,35,33,20

```
 P   Q   R
25   0   3    5,23,38, 4,26,40
 2   4   0    5,24
 1  19   8    5,24,21,57, 2,27,39,22,30
29   0  14    5,25, 9,13,16, 9,32,50,22,13,20
 0   1   8    5,25,31,15
52   6   0    5,25,46,52,24,10,39,35,25,26,24
35  13   0    5,26, 8,56,41, 4,21,30,14,24
18  20   0    5,26,31, 2,27,41,11, 2,24
 1  27   0    5,26,53, 9,44, 7,12,54
 2   0  18    5,27, 2,55,41,57,21,40
44   0   5    5,27,18,37,30, 8,23,42,13,20
17   2   0    5,27,40,48
 0   9   0    5,28, 3
 1  25   9    5,28,25,13,30,14,30, 7, 1,52,30
17   0   9    5,29,13, 5,11, 6,40
 0   7   9    5,29,35,23,26,15
33  18   0    5,30,13,33,23,35, 9,46,22, 4,48
16  25   0    5,30,35,55,44,31,56,55,40,48
 1  33   1    5,30,58,19,36,25,18,18,40,30
 0   5  18    5,31, 8,12,53,43,49,41,15
32   0   0    5,31,24, 6,28,16
15   7   0    5,31,46,33,36
 0  15   1    5,32, 9, 2,15
36   0  11    5,32,57,26,32,52, 0,59,15,33,20
 5   0   4    5,33,20
 0  13  10    5,33,42,34,58,49,41,15
14  30   0    5,34,43,52,41,20,20,53,22,23,36
 9   0  15    5,34,53,52,42,57,46,40
51   0   2    5,35, 9,57, 7,20,35,47,33,20
 0  11  19    5,35,16,34, 3,24, 7,33,30,56,15
30   5   0    5,35,32,39,33, 7,12
13  12   0    5,35,55,23,31,12
 0  21   2    5,36,18, 9, 1,41,15
24   0   6    5,37, 7, 9,37,46,40
 1   4   3    5,37,30
 0  19  11    5,37,52,51,55, 3,48,30,56,15
 1   2  12    5,39, 5, 3, 7,30
45   3   0    5,39,21,19,35,11, 6,14,24
28  10   0    5,39,44,19, 2,47, 2,24
11  17   0    5,40, 7,20, 3,50,24
 0  27   3    5,40,30,22,38,27,30,56,15
 1   0  21    5,40,40,33, 1,12,15, 4,10
43   0   8    5,40,56,54, 3,53,44,41,28,53,20
12   0   1    5,41,20
 1  10   4    5,41,43, 7,30
16   0  12    5,42,56, 7,54, 4,26,40
60   1   0    5,43,12,35,27,11,58,15, 5,48,48
 1   8  13    5,43,19,21,54,50,37,30
43   8   0    5,43,35,50,34,52,29,34, 4,48
26  15   0    5,43,59, 7,17, 4, 7,40,48
 9  22   0    5,44,22,25,33,53,16,48
 0  33   4    5,44,45,45,25,26,21,34,27,11,15
 1   6  22    5,44,56, 3,25,58, 9,15,28, 7,30
31   0   3    5,45,12,36,44,26,40
 8   4   0    5,45,36
 1  16   5    5,45,59,24,50,37,30
 4   0   7    5,47,13,20
 1  14  14    5,47,36,51,26,16,45,28, 7,30
41  13   0    5,47,53,32,27,48,38,56,15,21,36
24  20   0    5,48,17, 6,37,31,55,46,33,36
 7  27   0    5,48,40,42,23, 3,41,45,36
 8   0  18    5,48,51, 7,24,45,11, 6,40
50   0   5    5,49, 7,52, 0, 8,57,17, 2,13,20
23   2   0    5,49,31,31,12
 6   9   0    5,49,55,12
 1  22   6    5,50,18,54,24,15,28, 7,30
23   0   9    5,51, 9,57,31,51, 6,40
 0   4   6    5,51,33,45
22  25   0    5,52,38,19,27,30, 4,43,23,31,12
 5  32   0    5,53, 2,12,54,50,59,31,55,12
 0   2  15    5,53,12,45,45,18,45
38   0   0    5,53,29,42,54, 9, 4
21   7   0    5,53,53,39,50,24
 4  14   0    5,54,17,38,24
 1  28   0    5,54,41,38,35, 3,39,43,35,37,30
 0   0  24    5,54,52,14,23,45,15,41,50,25
11   0   4    5,55,33,20
 0  10   7    5,55,57,25,18,45
15   0  15    5,57,13,28,13,49,37,46,40
 3  37   0    5,57,26,59,34,32, 7,46,34, 8,24
57   0   2    5,57,30,36,55,49,58,10,43,33,20
 0   8  16    5,57,37,40,19,37,44, 3,45
36   5   0    5,57,54,50,11,19,40,48
19  12   0    5,58,19, 5, 5,16,48
 2  19   0    5,58,43,21,37,48
30   0   6    5,59,35,38,16,17,46,40
 1   1   0    6
 0  16   8    6, 0,24,23,22,44, 3,45
 3   0  10    6, 1,41,23,20
51   3   0    6, 1,58,44,53,31,50,39,21,36
 0  14  17    6, 2, 5,53,34,52,27,21,47,48,45
34  10   0    6, 2,23,16,18,58,10,33,36
17  17   0    6, 2,47,49,24, 5,45,36
 0  24   0    6, 3,12,24, 9, 1,21
 7   0  21    6, 3,23,15,13,17, 4, 4,26,40
18   0   1    6, 4, 5,20
 1   7   1    6, 4,30
 0  22   9    6, 4,54,41,40,16, 6,47,48,45
22   0  12    6, 5,47,52,25,40,44,26,40
 1   5  10    6, 6,12,39,22,30
```

```
11, 7,25, 9,57,39,22,30
11, 6,4C
11, 5,54,53, 5,44, 1,58,31, 6,40
11, 4,18, 4,30
11, 3,33, 7,12
11, 3, 1,17,53,43,11,54, 1,56,30,36, 2, 6,33,45
11, 2,16,25,47,27,39,22,30
11, 1,31,36,43,23, 1, 4,11,51, 6,40
11, 0,46,50,41,16,57,15,45,26,59,22, 1, 4,44,46,25,11, 6,40
11, 0,27, 6,47,10,19,32,44, 9,36
10,59,55,26,24, 6,30, 4,10,29,17,48,45
1C,59,10,46,52,30
10,58,26,10,22,13,20
10,57,41,36,53, 4,13,48, 9,59,10,37, 2,13,20
10,56, 6
10,55,21,36
10,54, 5,51,23,54,43,20
10,53,21,35,31,44,12,54,30,57,53,15, 3,42,13,20
10,52,37,22,39,17,29, 8,53,46,39,22,29,12,50, 8,48,34,40,39,30,22,13,20
10,52,17,53,22, 8,43, 0,28,48
10,51,46,36,26,46,25,15,14, 3,45
10,51, 2,30
10,50,18,26,32,19, 5,40,44,26,40
10,48,43,54, 4,55,18,45
1C,48
10,47,16, 8,53,20
1C,45,17,37,18,45, 9, 2,43,54,57, 2,16,59,28,43,27,24,26,40
10,44,56,21, 9,30,14,24
10,44,27,25,18,51,2C,55,38,22, 2,51,49,34,13, 7,30
10,44,14,42,2C,23,25,26,24
10,43,42,48,35,19,55,18,45
10,43, 0,14,48,53,20
10,42,16,43,59,19,35,58,45,22,38, 1,28,53,20
10,40,43,21,33,45
10,4C
1C,39,16,41,22,18,16,17,46,40
10,37, C,35,42,43,12
10,36,3C, 2,46,46,16,13,28,15,50,58,35,37,30
10,35,46,58,21,33,45
10,35, 3,56,51,14,53,49,37,46,40
10,34,20,58,15,37,52,34,19,37,54,35,32,14, 9,22,57,46,40
10,34, 2, 1,42,53, 6,45,49,35,36,57,36
10,33,31,37,2C,44,38,28, C,28, 7,30
1C,32,48,45
10,32, 5,55,33,20
10,29,51,21,36
10,29,21, 9,15, 7,57,28, 5, 7,28, 6,32,56,23, 7,47,34,41,15
10,29, 8,44, 9,36
10,28,38,33,51,22,44,10,2C,30,28, 7,30
1C,27,56, 1,20,33,2C
1C,27,13,31,42,28, 2,47,32, 7,34,19,15,13,20
10,26,31, 4,56,55,11,10,56,25,35,23,59,14,43,20,27,26, 5,25,55,33,20
1C,26,12,22,26, 3,34, 5,15,38,52,48
10,25,42,20,35,18, 9,50,37,30
1C,25
10,24,17,42,16,37,31,51, 6,40
10,22, 4,48
10,21,22,42, 8
10,20,52,54,10,44,40,39,50,37,30
10,20,10,53,1C,40,19,45,11, 6,40
10,19,28,55, 1,12, 8,41, 1,21,33, 9,23,30,41,58,31, 6,40
10,19,10,25, 6,43,45,24
10,18,4C,43,30, 6, 5,41,24,49,57,56,57,11,15
10,17,58,51,26,43, 7,30
10,17,17, 2,13,20
10,16,35,15,49,45,12,56,24,21,43,42,13,20
10,15, 5,37,30
10,14,24
10,12,31,29,33,30,12, 6, 6,31,46,10,22,13,20
10,11,5C, 2,29,20, 8,34,35,24,59,24,49,53,17, 0,45,32,30,37, 2,13,20
10,11,31,46,57,24, 6,40,19,12
1C,11, 2,26,40, 6, 1,10,31,56, 0,56,15
10,1C,21, 5,37,30
10, 9,39,47,22,47,54, 4,26,40
10, 8,58,31,55,48,21,40, 9,14,47,36,30,56,47,24,26,40
10, 8,4C,20,5C,46,11,17,35,36,35,28,53,45,36
10, 7,3C
1C, 6,48,53,20
10, 4,1C,49,51,56,11,26, 0,54,18,40,49,16,19,45,56, 5,26,32, 8, 7,14,34, 4,26,40
10, 4,10,42,28,55,38,22, 9,43,10,11, 5,13,19,48,16,52,30
10, 3,5£,47,11,36,57,36
10, 3,2£,49,18, 7,25,36,19,41,15
10, 2,48,58,53,20
10, 2, 8,11,14,22, 7,28,50, 2,28, 8,53,20
1C, 0,4C,38,57,53,26,15
10
9,59,19,23,47, 9,37,46,40
9,57,11,48,28,48
9,56,43,10, 6,20,52,42,37,44,51,32,25,53,54,22,30
9,56,31,23,38,52,48
9,56, 2,47,12,42,53,26,15
9,55,22,27, 3, 2,42,57,46,40
9,54,42, 9,37, 9,15,32,10,54,17,25,48,58,16,17,46,40
9,54,24,24, 6,27,17,35,27,44,38,24
9,53,15,42,11,15
9,52,35,33,20
9,51,55,27,11,45,48,25,20,59,15,33,20
9,50,26,24
9,49,45,26,24
```

P	Q	R	
49	8	0	6, 6,30,13,57,11,59,32,21, 7,12
32	15	0	6, 6,55, 3,46,12,24,11,31,12
15	22	0	6, 7,19,55,16, 8,49,55,12
0	30	1	6, 7,44,48,27, 8, 7, 0,45
1	3	19	6, 7,55,47,39,42, 1,52,30
37	0	3	6, 8,13,27,11,24,26,40
14	4	0	6, 8,38,24
1	13	2	6, 9, 3,22,30
10	0	7	6,10,22,13,20
1	11	11	6,10,47,18,52, 1,52,30
30	20	0	6,11,30,15, 4, 2, 3,29,39,50,24
13	27	0	6,11,55,25,12,35,56,32,38,24
14	0	18	6,12, 6,31,54,24,11,51, 6,40
0	36	2	6,12,20,37, 3,28,28, 6, 0,33,45
1	9	20	6,12,31,44,30,26,48,23,54,22,30
29	2	0	6,12,49,37,16,48
12	9	0	6,13,14,52,48
1	19	3	6,13,40,10, 1,52,30
29	0	9	6,14,34,37,21,58,31, 6,40
0	1	3	6,15
1	17	12	6,15,25,24,21,10,53,54,22,30
11	32	0	6,16,34,21,46,30,23,30, 2,52,48
2	0	13	6,16,46,36,48,20
44	0	0	6,17, 3,41,45,45,40,16
27	7	0	6,17,29,14,29,45,36
10	14	0	6,17,54,48,57,36
1	25	4	6,18,20,25, 9,23,54,22,30
6	0	24	6,18,31,43,21,20,16,44,37,46,40
17	0	4	6,19,15,33,20
0	7	4	6,19,41,15
21	0	15	6,21, 2,22, 6,44,56,17,46,40
0	5	13	6,21,28,11, 0,56,15
42	5	0	6,21,46,29,32, 4,59,31,12
25	12	0	6,22,12,21,25,37,55,12
8	19	0	6,22,38,15, 4,19,12
1	31	5	6,23, 4,10,28,15,57,18,16,52,30
0	3	22	6,23,15,37, 8,51,16,57,11,15
36	0	6	6,23,34, 0,49,22,57,46,40
7	1	0	6,24
0	13	5	6,24,26, 0,56,15
9	0	10	6,25,48, 8,53,20
57	3	0	6,26, 6,39,53, 5,58, 1,59, 2,24
0	11	14	6,26,14,17, 9,11,57,11,15
40	10	0	6,26,32,49,24,14, 3,15,50,24
23	17	0	6,26,59, 0,41,42, 8,38,24
6	24	0	6,27,25,13,45,37,26,24
13	0	21	6,27,36,48,14,10,12,20,44,26,40
24	0	1	6,28,21,41,20
5	6	0	6,28,48
0	19	6	6,29,14,20,26,57,11,15
28	0	12	6,30,11, 3,55,23,27,24,26,40
1	2	7	6,30,37,30
0	17	15	6,31, 3,57,52, 3,51, 9, 8,26,15
38	15	0	6,31,22,44, 1,17,13,48,17,16,48
21	22	0	6,31,49,14,57,13,25,14,52,48
4	29	0	6,32,15,47,40,56,39,28,48
1	0	16	6,32,27,30,50,20,50
43	0	3	6,32,46,21, 0,10, 4,26,40
20	4	0	6,33,12,57,36
3	11	0	6,33,39,36
0	25	7	6,34, 6,16,12,17,24, 8,26,15
16	0	7	6,35, 3,42,13,20
1	8	8	6,35,30,28, 7,30
19	27	0	6,36,43, 6,53,26,20,18,48,57,36
20	0	18	6,36,54,58, 2, 1,48,38,31, 6,40
2	34	0	6,37, 9,59,31,42,21,58,24,36
1	6	17	6,37,21,51,28,28,35,37,30
35	2	0	6,37,40,55,45,55,12
18	9	0	6,38, 7,52,19,12
1	16	0	6,38,34,50,42
35	0	9	6,39,32,55,51,26,25,11, 6,40
4	0	2	6,40
1	14	9	6,40,27, 5,58,35,37,30
8	0	13	6,41,52,39,15,33,20
0	39	0	6,42, 7,52, 1,21, 8,44,53,24,27
50	0	0	6,42,11,56,32,48,42,57, 4
1	12	18	6,42,19,52,52, 4,57, 4,13, 7,30
33	7	0	6,42,39,11,27,44,38,24
16	14	0	6,43, 6,28,13,26,24
1	22	1	6,43,33,46,50, 1,30
23	0	4	6,44,32,35,33,20
0	4	1	6,45
1	20	10	6,45,27,26,18, 4,34,13, 7,30
27	0	15	6,46,26,31,35,11,56, 2,57,46,40
0	2	10	6,46,54, 3,45
48	5	0	6,47,13,35,30,13,19,29,16,48
31	12	0	6,47,41,10,51,20,26,52,48
14	19	0	6,48, 8,48, 4,36,28,48
1	28	2	6,48,36,27,10, 9, 1, 7,30
0	0	19	6,48,48,39,37,26,42, 5
42	0	6	6,49, 8,16,52,40,29,37,46,40
13	1	0	6,49,36
0	10	2	6,50, 3,45
15	0	10	6,51,31,21,28,53,20
0	8	11	6,51,59,14,17,48,45
46	10	0	6,52,19, 0,41,50,59,28,53,45,36
29	17	0	6,52,46,56,44,28,57,12,57,36
12	24	0	6,52,14,54,40,39,56, 9,36
1	34	3	6,53,42,54,30,31,37,53,20,37,30
0	6	20	6,53,55,16, 7, 9,47, 6,33,45

9,49,21, 9,14,25, 3,54,41,43,33,52, 1,52,30
9,48,41,16,15,31,15
9,48, 1,25,58,33,47,37, 3,52, 5,55,33,20
9,47,21,38,23,21,44,14, 0,23,59,26,14,17,33, 7,55,43,12,35,33,20
9,47, 4, 6, 1,55,50,42,25,55,12
9,46,35,56,48, 5,46,43,42,39,22,30
9,45,56,15
9,45,16,35,53, 5,11, 6,40
9,43,12
9,42,32,32
9,41,25,12,21,15,18,31, 6,40
9,40,45,51,34,52,38, 8,27,31,27,20, 3,17,31,51, 6,40
9,40,28,31, 2,33,12,57,36
9,40, 6,33,28,15,32,34,34,28, 8,19,59,18, 4,34,29,50,49,28,26,59,45,11, 6,40
9,39,49,14, 6,21, 4,53,45,36
9,39,21,25,43,47,55,46,52,30
9,38,42,13,20
9,38, 3, 3,35,23,38,22,52,50,22,13,20
9,36,39, 1,24,22,30
9,36
9,35,21, 1,14, 4,26,40
9,33,35,39,50, 0, 8, 2,25,42,10,42, 1,46,12,11,57,41,43,42,13,20
9,23,18,32, 8,26,52,48
9,32,51, 2,30, 5,38,36, 7,26,15,52,44, 3,45
9,32,12,16,31,24,22,30
9,31,33,33,10, 7,24,26,40
9,30,54,52,26, 4, 5,18,53,40, 7, 7,59, 0,44,26,40
9,30,37,49,32,35,48, 5,14,38, 3,15,50,24
9,29,31,52,30
9,28,53,20
9,26,52,13,26,24
9,26,13,51,44,38,24
9,25,46,42,28,14,27,45,18,27,25,18,45
9,25, 8,25,12,30
9,24,30,10,32,13,14,30,46,54,48,53,20
9,23,51,58,27,13,40, 3,50,47, 1,51,35,19,15, 0,24,41,28,53,20
9,23,35, 8,11,27,12,40,44, 4,59,31,12
9,23, 8, 6,31,46,20,51,33,45
9,22,30
9,21,51,56, 2,57,46,40
9,19,52,19,12
9,19,25,28,13,27, 4,24,57,53,18,19, 9,16,47,13,35,37,30
9,19,14,25,55,12
9,18,47,36,45,40,12,35,51,33,45
9,18, 9,47,51,36,17,46,40
9,17,32, 1,31, 4,55,48,55,13,23,50,27, 9,37,46,40
9,17,15,22,36, 3, 5,14,29,45,36
9,16,10,58,18, 2,48,45
9,15,33,20
9,14,55,44,14,46,41,38,45,55,33,20
9,13,35, 3,45
9,12,57,36
9,12,20,10,47, 6,40
9,11,53,41,29,33, 2,48,45
9,11,16,20,36, 9,10,53,29,52,35,33,20
9,10,39, 2,14,24, 7,43, 7,52,29,28,20,53,57,18,40,59,15,33,20
9,10,22,35,39,18,36,17,16,48
9, 9,56,12, 0, 5,25, 3,28,44,24,50,37,30
9, 9,18,59, 3,45
9, 8,41,48,38,31, 6,40
9, 8, 4,40,44,13,31,30, 8,19,18,50,51,51, 6,40
9, 6,45
9, 6, 8
9, 4,27,59,36,26,50,45,25,48,14,22,33, 5,11, 6,40
9, 4,11,44, 6, 8,38,24
9, 3,51, 8,52,44,34,17,24,48,52,48,44,20,41,47,20,28,53,52,55,18,31, 6,40
9, 3,34,54,28,27,15,50,24
9, 3, 8,50,22,18,41, 2,41,43, 7,30
9, 2,32, 5
9, 1,55,22, 6,55,54,43,57, 2,13,20
9, 0,36,35, 4, 6, 5,37,30
9
8,59,23,27,24,26,40
8,57,28,37,37,55,12
8,57, 8,17,39,29,56,49,47,28,16,36,17, 7,50,54, 9,51,30,15,13,53, 6,16,57,17, 2,13,/
8,57, 2,51, 5,42,47,26,21,58,22,23,11,18,30,56,15   /20
8,56,52,15,16,59,31,12
8,56,26,30,29,26,36, 5,37,30
8,55,50,12,20,46,26,40
8,55,13,56,39,26,19,58,57,48,51,41,14, 4,26,40
8,53,56, 7,58, 7,30
8,53,20
8,52,43,54,28,35,13,34,48,53,20
8,51,26,27,36
8,50,50,29,45,36
8,50,25, 2,18,58,33,31,13,33,12,28,49,41,15
8,49,49, 8,37,58, 7,30
8,49,13,17,22,42,24,51,21,28,53,20
8,48,37,28,33, 1,33,48,36,21,35,29,36,51,47,49, 8, 8,53,20
8,48,21,41,25,44,15,38,11,19,40,48
8,47,56,21, 7,17,12, 3,20,23,26,15
8,47,20,37,30
8,46,44,56,17,46,40
8,44,52,48
8,44,17,16,48
8,43,52, 8,12,48,56,48,37, 5,23,26,15
8,43,16,41, 7, 7,46,40
8,42,41,16,25,23,22,19,36,46,18,36, 2,57,46,40
8,42, 5,54, 7,25,59,19, 7, 1,19,29,59,22,16, 7, 2,51,44,31,36,17,46,40
8,41,50,18,41,42,58,24,23, 2,24

P	Q	R	Value
30	0	1	6,54,15, 8, 5,20
11	6	0	6,54,43,12
0	16	3	6,55,11,17,48,45
34	0	12	6,56,11,48,11, 5, 1,14, 4,26,40
3	0	5	6,56,40
0	14	12	6,57, 8,13,43,32, 6,33,45
27	22	0	6,57,56,31,57, 2,18,55,52,19,12
10	29	0	6,58,24,50,51,40,26, 6,43,12
7	0	16	6,58,37,20,53,42,13,20
49	0	3	6,58,57,26,24,10,44,44,26,40
26	4	0	6,59,25,49,26,24
9	11	0	6,59,54,14,24
0	22	4	7, 0,22,41,17, 6,33,45
22	0	7	7, 1,23,57, 2,13,20
1	5	5	7, 1,52,30
0	20	13	7, 2,21, 4,53,49,45,38,40,18,45
8	34	0	7, 3,38,39,29,49,11,26,18,14,24
1	3	14	7, 3,51,18,54,22,30
41	2	0	7, 4,11,39,28,58,52,48
24	9	0	7, 4,40,23,48,28,48
7	16	0	7, 5, 9,10, 4,48
0	28	5	7, 5,37,58,18, 4,23,40,18,45
1	1	23	7, 5,50,41,16,30,18,50,12,30
41	0	9	7, 6,11, 7,34,52,10,51,51, 6,40
10	0	2	7, 6,40
1	11	6	7, 7, 8,54,22,30
14	0	13	7, 8,40, 9,52,35,33,20
56	0	0	7, 9, 0,44,18,59,57,48,52,16
1	9	15	7, 9, 9,12,23,33,16,52,30
39	7	0	7, 9,29,48,13,35,36,57,36
22	14	0	7, 9,58,54, 6,20, 9,36
5	21	0	7,10,28, 1,57,21,36
29	0	4	7,11,30,45,55,33,20
4	3	0	7,12
1	17	7	7,12,29,16, 3,16,52,30
2	0	8	7,14, 1,40
54	5	0	7,14,22,29,52,14,12,47,13,55,12
1	15	16	7,14,51, 4,17,50,56,50, 9,22,30
37	12	0	7,14,51,55,34,45,48,40,19,12
20	19	0	7,15,21,23,16,54,54,43,12
3	26	0	7,15,50,52,58,49,37,12
6	0	19	7,16, 3,54,15,56,28,53,20
48	0	6	7,16,24,50, 0,11,11,36,17,46,40
19	1	0	7,16,54,24
2	8	0	7,17,24
1	23	8	7,17,53,38, 0,19,20, 9,22,30
21	0	10	7,18,57,26,54,48,53,20
0	5	8	7,19,27,11,15
35	17	0	7,20,18, 4,31,26,53, 1,49,26,24
18	24	0	7,20,47,54,19,22,35,54,14,24
1	31	0	7,21,17,46, 8,33,44,24,54
0	3	17	7,21,30,57,11,38,26,15
36	0	1	7,21,52, 8,37,41,20
17	6	0	7,22,22, 4,48
0	13	0	7,22,52, 3
0	1	26	7,23,35,17,59,41,34,37,18, 1,15
9	0	5	7,24,26,40
0	11	9	7,24,56,46,38,26,15
16	29	0	7,26,18,30,15, 7, 7,51,10, 4,48
13	0	16	7,26,31,50,17,17, 2,13,20
1	37	1	7,26,48,44,28,10, 9,43,12,40,30
55	0	3	7,26,53,16, 9,47,27,43,24,26,40
0	9	18	7,27, 2, 5,24,32,10, 4,41,15
32	4	0	7,27,23,32,44, 9,36
15	11	0	7,27,53,51,21,36
0	19	1	7,28,24,12, 2,15
28	0	7	7,29,29,32,50,22,13,20
1	2	2	7,30
0	17	10	7,30,30,29,13,25, 4,41,15
1	0	11	7,32, 6,44,10
47	2	0	7,32,28,26, 6,54,48,19,12
30	9	0	7,32,59, 5,23,42,43,12
13	16	0	7,33,29,46,45, 7,12
0	25	2	7,34, 0,30,11,16,41,15
5	0	22	7,34,14, 4, 1,36,20, 5,33,20
16	0	2	7,35, 6,40
1	8	3	7,35,37,30
0	23	11	7,36, 8,22, 5,20, 8,29,45,56,15
20	0	13	7,37,14,50,32, 5,55,33,20
62	0	0	7,37,36,47,16,15,57,40, 7,45, 4
1	6	12	7,37,45,49,13, 7,30
45	7	0	7,38, 7,47,26,29,59,25,26,24
28	14	0	7,38,38,49,42,45,30,14,24
11	21	0	7,39, 9,54, 5,11, 2,24
0	31	3	7,39,41, 0,33,55, 8,45,56,15
1	4	21	7,39,54,44,34,37,32,20,37,30
35	0	4	7,40,16,48,59,15,33,20
10	3	0	7,40,48
1	14	4	7,41,19,13, 7,30
8	0	8	7,42,57,46,40
1	12	13	7,43,29, 8,35, 2,20,37,30
43	12	0	7,43,51,23,17, 4,51,55, 0,28,48
26	19	0	7,44,22,48,50, 2,34,22, 4,48
9	26	0	7,44,54,16,30,44,55,40,48
12	0	19	7,45, 8, 9,53, 0,14,48,53,20
25	1	0	7,46, 2, 1,36
8	8	0	7,46,33,36
1	20	5	7,47, 5,12,32,20,37,30
27	0	10	7,48,13,16,42,28, 8,53,20
0	2	5	7,48,45

8,41,25,17, 9,25, 8,12,11,15
8,40,50
8,40,14,45,13,51,16,32,35,33,20
8,38,59, 7,15,56,15
8,38,24
8,37,48,55, 6,40
8,36,45, 4,18,53,36,27,39,15,33,20
8,36,14, 5,51, 0, 7,14,11, 7,57,37,49,35,34,58,45,55,33,20
8,35,58,40,55,36,11,31,12
8,35,33,56,15, 5, 4,44,30,41,38,17,27,39,22,30
8,34,55, 2,52,15,56,15
8,34,24,11,51, 6,40
8,33,45,23,11,27,40,47, 0,18, 6,25,11, 6,40
8,32,34,41,15
8,32
8,31,25,21, 5,50,37, 2,13,20
8,29,51,42, 4,26,47, 8,49,30,49,30,41,34,24,10,37,57, 5,30,51,51, 6,40
8,29,36,28,34,10,33,36
8,29,12, 2,13,25, 0,58,46,36,40,46,52,30
8,28,37,34,41,15
8,28, 3, 9,28,59,55, 3,42,13,20
8,27,20,46,36,30,18, 3,27,42,19,40,25,47,19,30,22,13,20
8,27,13,37,22,18,29,24,39,40,29,34, 4,48
8,26,49,17,52,35,42,46,24,22,30
8,26,15
8,25,40,44,26,40
8,23,28,55,24, 6,21,58,28, 5,58,29,14,21, 6,30,14, 3,45
8,23,18,59,19,40,48
8,22,54,51, 5, 6,11,20,16,24,22,30
8,22,20,49, 4,26,40
8,21,46,49,21,58,26,14, 1,42, 3,27,24,26,40
8,20,33,52,28,14,31,52,30
8,20
8,19,26, 9,49,18, 1,28,53,20
8,17,39,50,24
8,17,15,58,25,17,23,55,31,27,22,57, 1,34,55,18,45
8,17, 6, 9,42,24
8,16,42,19,20,35,44,31,52,30
8,16, 8,42,32,32,15,48, 8,53,20
8,15,35, 8, 0,57,42,56,49, 5,14,31,30,48,33,34,48,53,20
8,15,20,20, 5,22,44,39,33, 7,12
8,14,50,34,48, 4,52,33, 7,51,58,21,33,45
8,14,23, 5, 9,22,30
8,13,49,37,46,40
8,13,16,12,39,48,10,21, 7,29,22,57,46,40
8,12, 4,30
8,11,31,12
8,10,34,23,32,56, 2,30
8,10, 1,11,38,48, 9,49,50,53,13,24,56,17,46,40
8, 9,28, 1,59,28, 6,51,40,19,59,31,51,54,37,36,36,26, 0,29,37,46,40
8, 9,13,25, 1,36,48, 8,53,20
8, 8,45,57,20, 4,48,56,25,32,48,45
8, 8,16,52,30
8, 7,43,49,54,14,19,15,33,20
8, 6,56,16,40,36,57, 2, 4,29,16,23, 7, 0,28,48
8, 6
8, 5,27, 6,40
8, 3,56,12,59, 3,51,47, 2,56,12,46,42,44,36,32,35,33,20
8, 3,43,45,52, 7,40,48
8, 3,25,27,53,32,57, 8,48,43,26,56,39,35, 3,48,44,52,21,13,42,29,47,39,15,33,20
8, 3,30,39,30,41,43,46,32, 8,52,10,39,50,37,30
8, 3,11, 1,45,17,34, 4,48
8, 2,47,51,26,29,56,29, 3,45
8, 2,15,11, 6,40
8, 1,42,32,59,29,41,59, 4, 1,58,31, 6,40
8, 0,32,31,10,18,45
8
7,59,27,31, 1,43,42,13,20
7,57,45,26,47, 2,24
7,57,22,32, 5, 4,42,10, 6,11,53,13,56,43, 7,30
7,56,50,13,46,10,18,45
7,56,17,57,38,26,10,22,13,20
7,55,45,43,41,43,24,25,44,43,25,56,39,10,37, 2,13,20
7,55,31,31,17, 9,50, 4,22,11,42,43,12
7,54,36,33,45
7,54, 4,26,40
7,53,32,21,45,24,38,44,16,47,24,26,40
7,52,23,31,12
7,52, 0,51,56,20,58, 6, 3,50,36, 4,54,42,17,20,50,41, 0,56,15
7,51,51,33, 7,12
7,51,28,55,23,32, 3, 7,45,22,51, 5,37,30
7,50,57, 1, 0,25
7,50,25, 8,46,51, 2, 5,39, 5,40,44,26,40
7,49,53,18,42,41,23,23,12,19,11,32,59,26, 2,30,20,34,34, 4,26,40
7,49,39,16,49,32,40,33,56,44, 9,36
7,49,16,45,26,28,37,22,58, 7,30
7,48,45
7,48,13,16,42,28, 8,53,20
7,46,33,36
7,46, 2, 1,36
7,45,39,40,38, 3,30,29,52,58, 7,30
7,45, 8, 9,53, 0,14,48,53,20
7,44,36,41,15,54, 6,30,46, 1, 9,52, 2,38, 1,28,53,20
7,44,22,48,50, 2,34,22, 4,48
7,43,29, 8,35, 2,20,37,30
7,42,57,46,40
7,42,26,26,52,18,54,42,18,16,17,46,40
7,41,19,13, 7,30
7,40,48

P	Q	R		
1	18	14	7,49,16,45,26,28,37,22,58, 7,3C	7,40,16,48,59,15,33,20
24	24	0	7,50,11, 5,56,40, 6,17,51,21,3€	7,39,23,37,1C, 7,39, 4,34,53,49,37,46,40
7	31	0	7,50,42,57,13, 7,59,22,33,36	7,3€,52,31,52, 0, 6,25,56,33,44,33,37,24,57,45,34, 9,22,57,46,40
0	0	14	7,50,57, 1, 0,25	7,38,38,49,42,45,3C,14,24
42	0	1	7,51,19,37,12,12, 5,20	7,38,16,50, 0, 4,30,52,53,57, 0,42,11,15
23	6	0	7,51,51,33, 7,12	7,37,45,49,13, 7,3C
6	13	0	7,52,23,31,12	7,37,14,50,32, 5,55,33,2C
1	26	6	7,52,55,31,26,44,52,58, 7,30	7,36,43,53,56,51,16,15, 6,56, 5,42,23,'2,35,33,20
4	0	25	7,53, 9,39,11,40,20,55,47,13,20	7,36,30,15,38, 4,38,28,11,42,26,36,40,19,12
15	0	5	7,54, 4,26,40	7,35,37,30
0	8	6	7,54,36,33,45	7,35, 6,40
19	0	16	7,56,17,57,38,26,10,22,13,20	7,33,29,46,45, 7,12
5	36	0	7,56,35,59,26, 2,50,22, 5,31,12	7,33,12,37,23,57, 8,34,30,40,44, 0,36,57,14,49,27, 4, 4,54, 6, 5,25,55,33,20
0	6	15	7,56,50,13,46,10,18,45	7,32,59, 5,23,42,43,12
38	4	0	7,57,13, 6,55, 6,14,24	7,32,37,21,5€,35,34,12,14,45,56,15
21	11	0	7,57,45,26,47, 2,24	7,32, 6,44,10
4	18	0	7,58,17,48,50,24	7,31,36, 8,25,46,35,36,37,31,51, 6,40
0	4	24	7,59, 4,31,26, 4, 6,11,29, 3,45	7,30,52, 6,33, 9,46, 8,35,15,59,36,57,36
34	0	7	7,59,27,31, 1,43,42,13,20	7,30,30,29,13,25, 4,41,15
3	0	0	8	7,30
0	14	7	8, 0,32,31,10,18,45	7,29,29,32,50,22,13,20
7	0	11	8, 2,15,11, 6,40	7,27,53,51,21,36
53	2	0	8, 2,38,19,51,22,27,32,28,48	7,27,32,22,34,45,39,31,58,18,38,39,19,25,25,46,52,30
0	12	16	8, 2,47,51,26,29,56,29, 3,45	7,27,23,32,44, 9,36
36	9	0	8, 3,11, 1,45,17,34, 4,48	7,27, 2, 5,24,32,10, 4,41,15
19	16	0	8, 3,43,45,52, 7,40,48	7,26,31,50,17,17, 2,13,20
2	23	0	8, 4,16,32,12, 1,48	7,26, 1,37,12,51,56,39, 8,10,43, 4,21,43,42,13,20
11	0	22	8, 4,31, 0,17,42,45,25,55,33,20	7,25,48,18, 4,50,28,11,35,48,28,48
22	0	2	8, 5,27, 6,40	7,24,56,46,38,26,15
1	5	0	8, 6	7,24,26,40
0	20	8	8, 6,32,55,33,41,29, 3,45	7,23,56,35,23,49,21,19, 0,44,26,40
26	0	13	8, 7,43,49,54,14,19,15,33,20	7,22,52, 3
1	3	9	8, 8,16,52,30	7,22,22, 4,48
51	7	0	8, 8,40,18,36,15,59,23, 8, 9,3€	7,22, C,51,55,48,47,56, 1,17,40,24, 1,24,22,30
34	14	0	8, 9,13,25, 1,36,32,15,21,36	7,21,3C,57,11,38,26,15
17	21	0	8, 9,46,33,41,31,46,33,36	7,21, 1, 4,28,55,20,42,47,54, 4,26,40
0	28	0	8,10,19,44,36,10,49,21	7,20,31,13,47,31,18,10,30,17,59,34,40,43, 9,50,56,47,24,26,40
1	1	18	8,10,34,23,32,56, 2,30	7,20,18, 4,31,26,53, 1,49,26,24
41	0	4	8,10,57,56,15,12,35,33,20	7,19,56,57,36, 4,20, 2,46,59,31,52,30
16	3	0	8,11,31,12	7,19,27,11,15
1	11	1	8,12, 4,30	7,18,57,26,54,48,53,20
0	26	9	8,12,37,50,15,21,45,10,32,48,45	7,18,27,44,35,22,49,12, 6,39,27, 4,41,28,53,20
14	0	8	8,13,49,37,46,40	7,17,24
1	9	10	8,14,23, 5, 9,22,30	7,16,54,24
32	19	0	8,15,20,20, 5,22,44,39,33, 7,12	7,16, 3,54,15,56,28,53,20
15	26	0	8,15,53,53,36,47,55,23,31,12	7,15,34,23,41, 9,28,36,2C,38,35,30, 2,28, 8,53,20
18	0	19	8,16, 8,42,32,32,15,48, 6,53,20	7,15,21,23,16,54,54,43,12
0	34	1	8,16,27,29,24,37,57,28, 0,45	7,15, 4,55, 6,11,39,25,55,51, 6,14,59,28,33,25,52,23, 7, 6,20,14,48,53,20
1	7	19	8,16,42,19,20,35,44,31,52,30	7,14,51,55,34,45,48,40,19,12
31	1	0	8,17, 6, 9,42,24	7,14,31, 4,17,50,56,50, 9,22,30
14	8	0	8,17,39,50,24	7,14, 1,40
1	17	2	8,18,13,33,22,30	7,13,32,17,41,32,43,47, 9,37,46,40
33	0	10	8,19,26, 9,49,18, 1,28,53,20	7,12,29,16, 3,16,52,30
2	0	3	8,20	7,12
1	15	11	8,20,33,52,28,14,31,52,30	7,11,3C,45,55,33,20
13	31	0	8,22, 5,49, 2, 0,31,20, 3,50,24	7,10,11,44,52,30, 6, 1,49,16,38, 1,31,19,39, 8,58,16,17,46,40
0	0	14	8,22,20,49, 4,26,40	7, 9,58,54, 6,20, 9,36
48	0	1	8,22,44,55,41, 0,53,41,20	7, 9,38,16,52,34,13,57, 5,34,41,54,33, 2,48,45
29	6	0	8,23,18,59,19,40,48	7, 9, 9,12,23,33,16,52,30
12	13	0	8,23,53, 5,16,48	7, 8,40, 9,52,35,33,20
1	23	3	8,24,27,13,32,31,52,30	7, 8,11, 9,19,33, 3,59,10,15, 5,20,59,15,33,20
21	0	5	8,25,40,44,26,40	7, 7, 8,54,22,30
0	5	3	8,26,15	7, 6,4C
1	21	12	8,26,49,17,52,35,42,46,24,22,20	7, 6,11, 7,34,52,10,51,51, 6,40
25	0	16	8,28, 3, 9,28,59,55, 3,42,13,20	7, 5, 5,10, 4,48
0	3	12	8,28,37,34,41,15	7, 4,40,23,48,28,48
44	4	0	8,29, 1,59,22,46,39,21,36	7, 4,20, 1,51,10,5C,48,58,50,33,59, 3,45
27	11	0	8,29,36,28,34,10,33,36	7, 3,51,18,54,22,30
10	18	0	8,30,11, 0, 5,45,36	7, 3,22,37,54, 9,55,53, 5,11, 6,40
1	29	4	8,30,45,233,57,41,16,24,22,30	7, 2,53,58,50,25,15, 2,53, 5,16,23,41,29,26,15,18,31, 6,40
0	1	21	8,31, 0,49,31,48,22,36,15	7, 2,41,21, 6,35,24,30,33, 3,44,38,24
40	0	7	8,31,25,21, 5,50,37, 2,13,20	7, 2,21, 4,53,49,45,38,40,18,45
9	0	0	8,32	7, 1,52,30
0	11	4	8,32,34,41,15	7, 1,23,57, 2,13,20
13	0	11	8,34,24,11,51, 6,40	6,59,54,14,24
59	2	0	8,34,48,53,10,47,57,22,38,43,12	6,59,34, 6,10, 5,18,18,43,24,58,44,21,57,35,25,11,43, 7,30
0	9	13	8,34,59, 2,52,15,56,15	6,59,25,49,26,24
42	9	0	8,35,23,45,52,18,44,21, 7,12	6,59, 5,42,34,15, 9,26,53,40,18,45
25	16	0	8,35,58,40,55,36,11,31,12	6,58,37,20,53,42,13,20
8	23	0	8,36,33,38,20,49,55,12	6,58, 9, 1, 8,18,41,51,41,25, 2,52,50,22,13,20
0	7	22	8,37,24, 5, 8,57,13,53,12,11,15	6,57,28,14,57,22,22,43,30,25,55,12
28	0	2	8,37,48,55, 6,40	6,57, 8,13,43,32, 6,33,45
7	5	0	8,38,24	6,56,40
0	17	5	8,38,59, 7,15,56,15	6,56,11,48,11, 5, 1,14, 4,26,40
32	0	13	8,40,14,45,13,51,16,32,35,33,20	6,55,11,17,48,45
1	0	6	8,40,50	6,54,43,12
0	15	14	8,41,25,17, 9,25, 8,12,11,15	6,54,15, 8, 5,20
40	14	0	8,41,50,18,41,42,58,24,23, 2,24	6,53,55,16, 7, 9,47, 6,33,45
23	21	0	8,42,25,39,56,17,53,39,50,24	6,53,27,15,27, 6,53,10, 7,24,26,40
6	28	0	8,43, 1, 3,34,35,32,38,24	6,52,59,16,40,48, 5,47,20,54,22, 6,15,40,27,59, 0,44,26,40
5	0	17	8,43,16,41, 7, 7,46,40	6,52,46,56,44,28,57,12,57,36
47	0	4	8,43,41,48, 0,13,25,55,33,20	6,52,27, 9, 0, 4, 3,47,36,33,18,37,58, 7,30
22	3	0	8,44,17,16,48	6,51,59,14,17,48,45
5	10	0	8,44,52,48	6,51,31,21,28,53,20
0	23	6	8,45,28,21,36,23,12,11,15	6,51, 3,30,33,10, 8,37,36,14,29, 8, 8,53,20
20	0	8	8,46,44,56,17,46,40	6,50, 2,45
1	6	7	8,47,20,37,30	6,49,36
21	26	0	8,48,57,29,11,15, 7, 5, 5,16,48	6,48,2C,59,42,20, 8, 4, 4,21,10,46,54,48,53,20
4	33	0	8,49,33,19,22,16,29,29,17,52,48	6,47,53,21,39,33,25,43, 3,36,39,36,33,15,31,20,30,21,40,24,41,28,53,20
1	4	16	8,49,49, 8,37,58, 7,30	6,47,41,10,51,2C,26,52,48

```
 P  Q  R
37  1  0   8,50,14,34,21,13,36                        6,47,21,37,46,44, 0,47, 1,17,20,37,30
20  8  0   8,50,50,29,45,36                           6,46,54, 3,45
 3 15  0   8,51,26,27,36                              6,46,26,31,35,11,56, 2,57,46,40
 0 29  7   8,52, 2,27,52,35,29,35,23,26,15            6,45,55, 1,17,12,14,26,46, 9,51,44,20,37,51,36,17,46,40
 1  2 25   8,52,18,21,35,37,53,32,45,37,30            6,45,46,53,53,5C,47,31,43,44,23,39,15,50,24
39  0 10   8,52,43,54,28,35,13,34,48,53,20            6,45,27,26,18, 4,34,13, 7,30
 8  0  3   8,53,20                                    6,45
 1 12  8   8,53,56, 7,58, 7,30                        6,44,32,35,33,20
12  0 14   8,55,50,12,20,44,26,40                     6,43, 6,28,13,26,24
 2 38  0   8,56,10,29,21,48,11,39,51,12,36            6,42,51,13,14,37,27,37,20,36,12,27,12,50,53,10,37,23,37,41,25,24,49,42,42,57,46,40
54  0  1   8,56,15,55,23,44,57,16, 5,20               6,42,47, 8,19,17, 5,34,46,28,46,47,23,28,53,12,11,15
 1 10 17   8,56,26,30,29,26,36, 5,37,30               6,42,39,11,27,44,38,24
35  6  0   8,56,52,15,16,59,31,12                     6,42,15,52,52, 4,57, 4,13, 7,30
18 13  0   8,57,28,37,37,55,12                        6,4C,27, 5,58,35,37,30
 1 20  0   8,58, 5, 2,26,42                           6,40
27  0  5   8,59,23,27,24,26,40                        6,39,32,55,51,26,25,11, 6,40
 0  2  0   9                                          6,38, 7,52,19,12
 1 18  9   9, 0,36,35, 4, 6, 5,37,30                  6,37,48,46,44,13,55, 8,25, 9,54,21,37,15,56,15
 0  0  9   9, 2,32, 5                                 6,37,21,51,28,28,35,37,30
50  4  0   9, 2,58, 7,20,17,45,59, 2,24               6,36,54,58, 2, 1,48,38,31, 6,40
33 11  0   9, 3,34,54,28,27,15,50,24                  6,36,28, 6,24,46,10,21,27,16,11,37,12,38,50,51,51, 6,40
16 18  0   9, 4,11,44, 6, 8,38,24                     6,36,16,16, 4,18,11,43,38,29,45,36
 1 26  1   9, 4,48,36,13,32, 1,30                     6,35,57,15,50,27,54, 2,30,17,34,41,15
 4  0 20   9, 5, 4,52,49,55,36, 6,40                  6,35,30,28, 7,30
46  0  7   9, 5,31, 2,30,13,59,30,22,13,20            6,35, 3,42,13,20
15  0  0   9, 6, 8                                    6,34,36,58, 7,50,32,16,53,59,30,22,13,20
 0  8  1   9, 6,45                                    6,23,39,36
 1 24 10   9, 7,22, 2,30,24,10,11,43, 7,30            6,33,12,57,36
19  0 11   9, 8,41,48,38,31, 6,40                     6,32,54, 6, 9,36,42,36,27,49, 2,34,41,15
 0  6 10   9, 9,18,59, 3,45                           6,32,27,30,50,20,50
48  9  0   9, 9,45,20,55,47,59,18,31,40,48            6,32, C,57,19, 2,31,44,42,34,43,57, 2,13,20
31 16  0   9,10,22,35,39,18,36,17,16,48               6,31,34,25,35,34,29,20,15,59,37,29,?1,42, 5,17, 8,48,23,42,13,20
14 23  0   9,10,59,52,54,13,14,52,48                  6,31,22,44, 1,17,13,48,17,16,48
 1 32  2   9,11,37,12,40,42,10,31, 7,30               6,31, 3,57,52, 3,51, 9, 8,26,15
 0  4 19   9,11,53,41,29,33, 2,48,45                  6,30,37,30
34  0  2   9,12,20,10,47, 6,40                        6,30,11, 3,55,23,27,24,26,40
13  5  0   9,12,57,36                                 6,28,48
 0 14  2   9,13,35, 3,45                              6,28,21,41,20
 7  0  6   9,15,33,20                                 6,27,36,48,14,10,12,20,44,26,40
 0 12 11   9,16,10,58,18, 2,48,45                     6,27,1C,34,23,15, 5,25,38,20,58,13,22,11,41,14, 4,26,40
29 21  0   9,17,15,22,36, 3, 5,14,29,45,36            6,26,59, 0,41,42, 8,38,24
12 28  0   9,17,53, 7,48,53,54,48,57,36               6,26,40,27,11,18,48,33,23, 1,13,43, 5,44,31,52,30
11  0 17   9,18, 9,47,51,36,17,46,40                  6,26,32,49,24,14, 3,15,50,24
53  0  4   9,18,36,35,12,14,19,39,15,33,20            6,26,14,17, 9,11,57,11,15
 0 10 20   9,18,47,36,45,40,12,35,51,23,45            6,25,48, 8,53,20
28  3  0   9,19,14,25,55,12                           6,25,22, 2,23,35,45,35,15,13,34,48,53,20
11 10  0   9,19,52,19,12                              6,24,26, 0,56,15
 0 20  3   9,20,30,15, 2,48,45                        6,24
26  0  8   9,21,51,56, 2,57,46,40                     6,23,34, 0,49,22,57,46,4C
 1  3  4   9,22,30                                    6,22,23,46,33,20, 5,21,37, 8, 7, 8, 1,10,48, 7,58,27,49, 8, 8,53,20
 0 18 12   9,23, 8, 6,31,46,20,51,33,45               6,22,12,21,25,37,55,12
10 33  0   9,24,51,32,39,45,35,15, 4,19,12            6,21,54, 1,40, 3,45,44, 4,57,30,35, 9,22,30
 1  1 13   9,25, 8,25,12,30                           6,21,28,11, 0,56,15
43  1  0   9,25,35,32,38,38,30,24                     6,21, 2,22, 6,44,56,17,46,40
26  8  0   9,26,13,51,44,38,24                        6,20,36,34,57,22,43,32,35,46,44,45,19,20,29,37,46,40
 9 15  0   9,26,52,13,26,24                           6,20,25,13, 1,43,52, 3,29,45,22,10,33,36
 0 26  4   9,27,30,37,44, 5,51,33,45                  6,19,41,15
 3  0 23   9,27,47,35, 2, 0,25, 6,56,40               6,19,15,33,20
14  0  3   9,28,53,20                                 6,17,54,48,57,36
 1  9  5   9,29,31,52,30                              6,17,36,41,33, 4,46,28,51, 4,28,51,55,45,49,52,40,32,48,45
18  0 14   9,31,33,33,10, 7,24,26,40                  6,17,29,14,29,45,36
60  0  1   9,32, 0,59, 5,19,57, 5, 9,41,20            6,17,11, 8,18,49,38,30,12,18,16,52,30
 1  7 14   9,32,12,16,31,24,22,30                     6,16,45,36,48,20
41  6  0   9,32,39,44,18, 7,29,16,48                  6,16,20, 7, 1,28,49,40,31,16,32,35,33,20
24 13  0   9,33,18,32, 8,26,52,48                     6,15,54,38,58, 9, 6,42,33,51,21,14,23,32,50, 0,16,27,39,15,33,20
 7 20  0   9,33,57,22,36,28,48                        6,15,43,25,27,38, 8,27, 9,23,19,40,48
 0 32  5   9,34,36,15,42,23,55,57,25,18,45            6,15,25,24,21,10,53,54,22,30
 1  5 23   9,34,53,25,43,16,55,25,46,52,30            6,15
33  0  5   9,35,21, 1,14, 4,26,40                     6,14,34,37,21,58,31, 6,40
 6  2  0   9,36                                       6,13,14,52,48
 1 15  6   9,36,39, 1,24,22,30                        6,12,56,58,48,58, 2,56,38,35,32,12,46,11,11,29, 3,45
 6  0  9   9,38,42,13,20                              6,12,49,37,16,48
56  4  0   9,39, 9,59,49,38,57, 2,58,33,36            6,12,31,44,30,26,48,23,54,22,30
 1 13 15   9,39,21,25,43,47,55,46,52,30               6,12, 6,31,54,24,11,51, 6,40
39 11  0   9,39,49,14, 6,21, 4,53,45,36               6,11,41,21, 0,43,17,12,36,48,55,53,38, 6,25,11, 6,40
22 18  0   9,40,28,31, 2,33,12,57,36                  6,11,30,15, 4, 2, 3,29,39,50,24
 5 25  0   9,41, 7,50,38,26, 9,36                     6,10,47,18,52, 1,52,30
10  0 20   9,41,25,12,21,15,18,31, 6,40               6,10,22,13,20
21  0  0   9,42,32,32                                 6, 9,57, 9,29,51, 7,45,50,37, 2,13,20
 4  7  0   9,43,12                                    6, 9, 3,22,30
 1 21  7   9,43,51,30,40,25,46,52,30                  6, 8,38,24
25  0 11   9,45,16,35,53, 5,11, 6,40                  6, 7,55,47,39,42, 1,52,30
 0  3  7   9,45,56,15                                 6, 7,30,53,44, 6, 7,15,39,55, 3,42,13,20
37 16  0   9,47, 4, 6, 1,55,50,42,25,55,12            6, 7, 6, 1,29,36, 5, 8,45,14,59,38,53,55,58,12,27,19,30,22,13,20
20 23  0   9,47,43,52,25,50, 7,52,19,12               6, 6,55, 3,46,12,24,11,31,12
 3 30  0   9,48,23,41,31,24,59,13,12                  6, 6,37,28, 0, 3,36,42,19, 9,36,33,45
 0  1 16   9,48,41,16,15,31,15                        6, 6,12,39,22,30
40  0  2   9,49, 9,31,30,15, 6,40                     6, 5,47,52,25,40,44,26,40
19  5  0   9,49,49,26,24                              6, 5,23, 7, 9,29, 1, 0, 5,32,52,33,54,34, 4,26,40
 2 12  0   9,50,29,24                                 6, 5,12,12,30,27,42,46,33,21,57,17,20,15,21,36
 1 27  8   9,51, 9,24,18,26, 6,12,39,22,30            6, 4,30
 2  0 26   9,51,27, 3,59,35,26, 9,44, 1,40            6, 4, 5,20
13  0  6   9,52,35,33,20                              6, 2,58,39,44,17,53,50,17,12, 9,35, 2, 3,27,24,26,40
 0  9  8   9,53,15,42,11,15                           6, 2,47,49,24, 5,45,36
18 28  0   9,55, 4,40,20, 9,30,28,13,26,24            6, 2,34, 5,55, 9,42,51,36,32,35,12,29,33,47,51,33,39,15,55,16,52,20,44,26,40
17  0 17   9,55,22,27, 3, 2,42,57,46,40               6, 2,23,16,18,58,10,33,36
 1 35  0   9,55,44,59,17,33,32,57,36,54               6, 2, 5,53,34,52,27,21,47,48,45
 0  7 17   9,56, 2,47,12,42,53,26,15                  6, 1,41,23,20
34  3  0   9,56,31,23,38,52,48
17 10  0   9,57,11,48,28,48
```

P	Q	R	
0	17	0	9,57,52,16, 3
32	0	8	9,59,19,23,47, 9,37,46,4C
1	0	1	10
0	15	9	10, 0,40,38,57,53,26,15
5	0	12	10, 2,48,58,53,20
49	1	0	10, 3,17,54,49,13, 4,25,3E
0	13	18	10, 3,29,49,18, 7,25,36,19,41,15
32	8	0	10, 3,58,47,11,36,57,36
15	15	0	10, 4,39,42,20, 9,36
0	23	1	10, 5,20,40,15, 2,15
9	0	23	10, 5,38,45,22, 8,26,47,24,26,40
20	0	3	10, 6,48,53,20
1	6	2	10, 7,30
0	21	10	10, 8,11, 9,27, 6,51,19,41,15
24	0	14	10, 9,39,47,22,47,54, 4,26,40
1	4	11	10,10,21, 5,37,30
47	6	0	10,10,50,23,15,19,59,13,55,12
30	13	0	10,11,31,46,17, 0,40,19,12
13	20	0	10,12,13,12, 6,54,43,12
0	29	2	10,12,54,40,45,13,31,41,15
1	2	20	10,13,12,59,26,10, 3, 7,30
39	0	5	10,13,42,25,19, 0,44,26,40
12	2	0	10,14,24
1	12	3	10,15, 5,37,30
12	0	9	10,17,17, 2,13,20
1	10	12	10,17,58,51,26,43, 7,30
45	11	0	10,18,28,31, 2,46,29,13,20,38,24
28	18	0	10,19,10,25, 6,43,25,49,26,24
11	25	0	10,19,52,22, 0,59,54,14,24
16	0	20	10,20,10,53,10,40,19,45,11, 6,4C
0	35	3	10,20,34,21,45,47,26,50, 0,56,15
1	8	21	10,20,52,54,10,44,40,39,50,37,30
27	0	0	10,21,22,42, 8
10	7	0	10,22, 4,48
1	18	4	10,22,46,56,43, 7,30
31	0	11	10,24,17,42,16,37,31,51, 6,40
0	0	4	10,25
1	16	13	10,25,42,20,35,18, 9,50,37,30
26	23	0	10,26,54,47,55,33,28,23,48,28,48
9	30	0	10,27,37,16,17,30,39,10, 4,48
4	0	15	10,27,56, 1,20,33,20
46	0	2	10,28,26, 9,36,16, 7, 6,40
25	5	0	10,29, 8,44, 9,36
8	12	0	10,29,51,21,36
1	24	5	10,30,34, 1,55,39,50,37,30
19	0	6	10,32, 5,55,33,20
0	6	5	10,32,42,48,45
23	0	17	10,35, 3,56,51,14,53,49,37,46,4C
7	35	0	10,35,27,59,14,43,47, 9,27,21,36
0	4	14	10,35,46,58,21,33,45
40	3	0	10,36,17,29,13,28,19,12
23	10	0	10,37, 0,35,42,43,12
6	17	0	10,37,43,45, 7,12
1	30	6	10,38,26,57,27, 6,35,30,28, 7,30
0	2	23	10,38,46, 1,54,45,28,15,18,45
38	0	8	10,39,16,41,22,18,16,17,46,40
7	0	1	10,40
0	12	6	10,40,43,21,33,45
11	0	12	10,43, 0,14,48,53,20
55	1	0	10,43,31, 6,28,29,56,43,18,24
0	10	15	10,43,43,48,35,19,55,18,45
38	8	0	10,44,14,42,20,23,25,26,24
21	15	0	10,44,58,21, 9,30,14,24
4	22	0	10,45,42, 2,56, 2,24
26	0	3	10,47,16, 8,53,20
3	4	0	10,48
0	18	7	10,48,43,54, 4,55,18,45
30	0	14	10,50,18,26,32,19, 5,40,44,26,4C
1	1	8	10,51, 2,30
53	6	0	10,51,33,44,48,21,19,10,50,52,48
0	16	16	10,51,46,36,26,46,25,15,14, 3,45
36	13	0	10,52,17,53,22, 8,43, 0,28,48
19	20	0	10,53, 2, 4,55,22,22, 4,48
2	27	0	10,53,46,19,28,14,25,48
3	0	18	10,54, 5,51,23,54,43,20
45	0	5	10,54,37,15, 0,16,47,24,26,40
18	2	0	10,55,21,36
1	9	0	10,56, 6
0	24	8	10,56,50,27, 0,29, 0,14, 3,45
18	0	9	10,58,26,10,22,13,20
1	7	9	10,59,10,46,52,30
34	18	0	11, 0,27, 6,47,10,19,32,44, 9,36
17	25	0	11, 1,11,51,29, 3,53,51,21,36
0	32	0	11, 1,56,39,12,50,36,37,21
1	5	18	11, 2,16,25,47,27,39,22,30
33	0	0	11, 2,48,12,56,32
16	7	0	11, 3,33, 7,12
1	15	1	11, 4,18, 4,30
37	0	11	11, 5,54,53, 5,44, 1,58,31, 6,40
6	0	4	11, 6,40
1	13	10	11, 7,25, 9,57,39,22,30
15	30	0	11, 9,27,45,22,40,41,46,45, 7,12
10	0	15	11, 9,47,45,25,55,33,20
0	38	1	11,10,13, 6,42,15,14,34,49, 0,45
52	0	2	11,10,19,54,14,41,11,35, 6,40
1	11	19	11,10,33, 8, 6,48,15, 7, 1,52,30
31	5	0	11,11, 5,19, 6,14,24
14	12	0	11,11,50,47, 2,24
1	21	2	11,12,36,18, 3,22,30
25	0	6	11,14,14,19,15,33,20

6, 1,1E,54,44,37,16,29,18, 1,28,53,20
6, 0,24,23,22,44, 3,45
6
5,59,35,38,16,17,46,40
5,58,19, 5, 5,16,48
5,58, 1,54, 3,48,31,37,34,38,54,55,27,32,20,37,30
5,57,54,50,11,19,40,48
5,57,37,40,19,37,44, 3,45
5,57,13,28,13,49,37,46,40
5,56,45,17,46,17,33,19,18,32,34,27,29,22,57,46,40
5,56,38,38,27,52,22,33,16,38,47, 2,24
5,55,57,25,18,45
5,55,33,20
5,55, 9,16,19, 3,29, 3,12,35,33,20
5,54,17,38,24
5,53,53,39,50,24
5,53,36,41,32,39, 2,20,49, 2, 8,19,13, 7,30
5,53,12,45,45,18,45
5,52,48,51,35, 8,16,34,14,19,15,33,20
5,52,24,59, 2, 1, 2,32,24,14,23,39,44,34,31,52,45,25,55,33,20
5,52,14,27,37, 9,30,25,27,33, 7,12
5,51,57,34, 4,51,28, 2,13,35,37,30
5,51,33,45
5,51, 9,57,31,51, 6,40
5,49,31,31,12
5,49,14,45,28,32,37,52,24,43,35,37,30
5,48,51, 7,24,45,11, 6,40
5,48,27,30,56,55,34,53, 4,30,52,24, 1,58,31, 6,40
5,48,17, 6,37,31,55,46,33,36
5,48, 3,56, 4,57,19,32,44,40,52,59,59,34,50,44,41,54,29,41, 4,11,51, 6,40
5,47,53,32,27,48,38,56,15,21,36
5,47,36,51,26,16,45,28, 7,30
5,47,13,20
5,46,49,50, 9,14,11, 1,43,42,13,20
5,45,59,24,50,37,30
5,45,3E
5,45,12,36,44,26,4C
5,44,32,42,52,35,44,18,26,10,22,13,20
5,44, 9,23,54, 0, 4,49,27,25,18,25,13, 3,43,19,10,37, 2,13,20
5,43,59, 7,17, 4, 7,40,48
5,43,42,37,30, 3,23, 9,40,27,45,31,38,26,15
5,43,19,21,54,50,37,30
5,42,56, 7,54, 4,26,40
5,42,32,55,27,38,27,11,20,12, 4,16,47,24,26,40
5,41,43, 7,30
5,40,20
5,40, 7,20, 3,50,24
5,39,54,28, 2,57,51,25,53, 0,33, 0,27,42,56, 7, 5,18, 3,40,34,34, 4,26,40
5,39,44,19, 2,47, 2,24
5,39,28, 1,28,56,40,39,11, 4,27,11,15
5,39, 5, 3, 7,30
5,38,42, 6,19,19,56,42,28, 8,53,20
5,38,19,11, 4,20,12, 2,18,28,13, 6,57,11,33, 0,14,48,53,20
5,38, 9, 4,54,52,19,36,26,26,59,42,43,12
5,37,52,51,55, 3,48,30,56,15
5,37,3C
5,37, 7, 9,37,46,4C
5,35,55,23,31,12
5,35,39,16,56, 4,14,38,58,43,58,59,29,34, 4,20, 9,22,30
5,35,32,39,33, 7,12
5,35,16,34, 3,24, 7,33,3C,56,15
5,34,53,52,42,57,46,33,36
5,34,31,12,54,38,57,29,21, 8, 2,18,16,17,46,40
5,33,42,34,58,49,41,15
5,33,20
5,32,57,26,32,52, C,59,15,33,20
5,32, 9, 2,15
5,31,46,33,36
5,31,3C,38,56,51,35,57, 0,58,15,18, 1, 3,16,52,30
5,31,24, 6,28,16
5,31, 8,12,53,43,49,41,15
5,30,45,48,21,41,30,32, 5,55,33,20
5,30,23,25,20,38,28,37,52,43,29,41, 0,32,22,23,12,35,33,20
5,30,13,33,23,35, 9,46,22, 4,48
5,29,57,43,12, 3,15, 2, 5,14,38,54,22,?0
5,29,35,23,26,15
5,29,13, 5,11, 6,4C
5,28,50,48,26,32, 6,54, 4,59,35,18,31, 6,40
5,28, 3
5,27,4C,48
5,27, 2,55,41,57,21,40
5,26,40,47,45,52, 6,27,15,28,56,37,31,51, 6,40
5,26,18,41,19,38,44,34,26,53,19,41,14,36,25, 4,24,17,20,19,45,11, 6,40
5,26, 8,56,41, 4,21,30,14,24
5,25,53,18,13,23,12,37,37, 1,52,30
5,25,31,15
5,25, 5,13,16, 9,32,50,22,13,20
5,24,21,57, 2,27,39,22,30
5,24
5,23,38, 4,26,4C
5,22,32,48,39,22,34,31,21,57,28,31, 8,29,44,21,43,42,13,20
5,22,2C,10,34,45, 7,12
5,22,16,58,35,41,58, 5,52,28,57,57,46,16,42,32,29,54,54, 9, 8,19,51,46,10,22,13,20
5,22,13,42,39,25,40,27,49,11, 1,25,54,47, 6,33,45
5,22, 7,21,10,11,42,43,12
5,21,51,54,17,39,57,39,22,30
5,21,30, 7,24,26,40
5,21, 8,21,59,39,47,59,22,41,19, 0,44,76,40
5,20,21,40,46,52,30

```
P   Q   R
 0   3   2   11,15
 1  19  11   11,15,45,43,50, 7,37, 1,52,30
 0   1  11   11,18,10, 6,15
46   3   0   11,18,42,39,10,22,12,28,48
29  10   0   11,19,28,38, 5,34, 4,48
12  17   0   11,20,14,40, 7,40,48
 1  27   3   11,21, 0,45,16,55, 1,52,30
 2   0  21   11,21,21, 6, 2,24,30, 8,20
44   0   8   11,21,53,48, 7,47,29,22,57,46,40
13   0   1   11,22,40
 0   9   3   11,23,26,15
17   0  12   11,25,52,15,48, 8,53,20
61   1   0   11,26,25,10,54,23,56,30,11,37,36
 0   7  12   11,26,38,43,49,41,15
44   8   0   11,27,11,41, 9,44,59, 8, 9,36
27  15   0   11,27,58,14,34, 8,15,21,36
10  22   0   11,28,44,51, 7,46,33,36
 1  33   4   11,29,31,30,50,52,43, 8,54,22,30
 0   5  21   11,29,52, 6,51,56,18,30,56,15
32   0   3   11,30,25,13,28,53,20
 9   4   0   11,31,12
 0  15   4   11,31,58,49,41,15
 5   0   7   11,34,26,40
 0  13  13   11,35,13,42,52,33,30,56,15
42  13   0   11,35,47, 4,55,37,17,52,30,43,12
25  20   0   11,36,34,13,15, 3,51,33, 7,12
 8  27   0   11,37,21,24,46, 7,23,31,12
 9   0  18   11,37,42,14,49,30,22,13,20
51   0   5   11,38,15,44, 0,17,54,34, 4,26,4C
24   2   0   11,39, 3, 2,24
 7   9   0   11,39,50,24
 0  21   5   11,40,37,48,48,30,56,15
24   0   9   11,42,19,55, 3,42,13,20
 1   4   6   11,43, 7,30
 0  19  14   11,43,55, 8, 9,42,56, 4,27,11,15
23  25   0   11,45,16,38,55, 0, 9,26,47, 2,24
 6  32   0   11,46, 4,25,49,41,59, 3,50,24
 1   2  15   11,46,25,31,30,37,30
39   0   0   11,46,59,25,48,18, 8
22   7   0   11,47,47,19,40,48
 5  14   0   11,48,35,16,48
 0  27   6   11,49,23,17,10, 7,19,27,11,15
 1   0  24   11,49,44,28,47,30,31,23,40,50
12   0   4   11,51, 6,40
 1  10   7   11,51,54,50,37,30
16   0  15   11,54,26,56,27,39,15,33,20
 4  37   0   11,54,53,59, 9, 4,15,33, 8,16,48
58   0   2   11,55, 1,13,51,39,56,21,27, 6,40
 1   8  16   11,55,15,20,39,15,28, 7,30
37   5   0   11,55,49,40,22,39,21,36
20  12   0   11,56,38,10,10,33,36
 3  19   0   11,57,26,43,15,36
31   0   6   11,59,11,16,32,35,33,20
 2   1   0   12
 1  16   8   12, 0,48,46,45,28, 7,30
 4   0  10   12, 3,22,46,40
52   3   0   12, 3,57,29,47, 3,41,18,43,12
 1  14  17   12, 4,11,47, 9,44,54,43,35,37,30
35  10   0   12, 4,46,32,37,56,21, 7,12
18  17   0   12, 5,35,38,48,11,31,12
 1  24   0   12, 6,24,48,18, 2,42
 8   0  21   12, 6,46,30,26,34, 8, 8,53,20
19   0   1   12, 8,10,40
 0   6   0   12, 9
 1  22   9   12, 9,49,23,20,32,13,35,37,30
23   0  12   12,11,35,44,51,21,28,53,20
 0   4   9   12,12,25,18,45
50   8   0   12,13, 0,27,54,23,59, 4,42,14,24
33  15   0   12,13,50, 7,32,24,48,23, 2,24
16  22   0   12,14,39,50,32,17,39,50,24
 1  30   1   12,15,29,36,54,16,14, 1,30
 0   2  18   12,15,51,35,19,24, 3,45
38   0   3   12,16,26,54,22,48,53,20
15   4   0   12,17,16,48
 0  12   1   12,18, 6,45
 0   0  27   12,19,18,49,59,29,17,42,10, 2, 5
11   0   7   12,20,44,26,40
 0  10  10   12,21,34,37,44, 3,45
31  20   0   12,23, 0,30, 8, 4, 6,59,19,40,48
14  27   0   12,23,50,50,25,11,53, 5,16,48
15   0  18   12,24,13, 3,48,48,23,42,13,20
 1  36   2   12,24,41,14, 6,56,56,12, 1, 7,3C
 0   8  19   12,25, 3,29, 0,53,36,47,48,45
30   2   0   12,25,39,14,33,36
13   9   0   12,26,29,45,36
 0  18   2   12,27,20,20, 3,45
30   0   9   12,29, 9,14,43,57, 2,13,20
 1   1   3   12,30
 0  16  11   12,30,50,48,42,21,47,48,45
12  32   0   12,33, 8,43,33, 0,47, 0, 5,45,36
 3   0  13   12,33,31,13,36,40
45   0   0   12,34, 7,23,31,31,20,32
28   7   0   12,34,58,28,59,31,12
11  14   0   12,35,49,37,55,12
 0  24   3   12,36,40,50,18,47,48,45
 7   0  24   12,37, 3,26,42,40,33,29,15,33,20
18   0   4   12,38,31, 6,40
 1   7   4   12,39,22,30
 0  22  12   12,40,13,56,48,53,34, 9,36,23,45
22   0  15   12,42, 4,44,13,29,52,35,33,20
```

```
5,20
5,19,38,20,41, 9, 8, 8,53,20
5,18,3C,17,51,21,36
5,18,15, 1,23,23, 8, 6,44, 7,55,29,17,48,45
5,17,53,29,10,46,52,30
5,17,31,58,25,37,26,54,48,53,20
5,17,1C,29, 7,48,56,17, 9,48,57,17,46, 7, 4,41,28,53,20
5,17, 1, 0,51,26,33,22,54,47,48,28,48
5,16,45,48,40,22,19,14, 0,14, 3,45
5,16,24,22,30
5,16, 2,57,46,40
5,14,55,40,48
5,14,4C,34,37,33,58,44, 2,33,44, 3,16,28,11,33,53,47,20,37,30
5,14,34,22, 4,48
5,14,19,16,55,41,22, 5,10,15,14, 3,45
5,13,58, 0,40,16,40
5,13,36,45,51,14, 1,23,46, 3,47, 9,37,46,40
5,13,15,32,28,27,35,35,28,12,47,41,59,37,21,40,13,43, 2,42,57,46,40
5,13, 6,11,13, 1,47, 2,37,49,26,24
5,12,51,10,17,39, 4,55,18,45
5,12,30
5,12, 8,51, 8,18,45,55,33,20
5,11, 2,24
5,10,41,21, 4
5,10,26,27, 5,22,2C,19,55,18,45
5,10, 5,26,35,20, 9,52,35,33,20
5, 9,44,27,30,36, 4,20,3C,40,46,34,41,45,20,59,15,33,20
5, 9,35,12,33,21,42,54,43,12
5, 9,20,21,45, 3, 2,50,42,24,58,58,28,35,37,30
5, 8,59,25,43,21,33,45
5, 8,38,31, 6,40
5, 8,17,37,54,52,36,28,12,10,51,51, 6,40
5, 7,32,48,45
5, 7,12
5, 6,51,12,39,30,22,13,2C
5, 6,15,44,46,45, 6, 3, 3,15,53, 5,11, 6,40
5, 5,55, 1,14,40, 4,17,17,42,29,42,24,56,38,30,22,46,15,18,31, 6,40
5, 5,45,53, 8,30,20, 9,36
5, 5,31,13,20, 3, 0,35,15,58, 0,28, 7,30
5, 5,10,32,48,45
5, 4,49,53,41,23,57, 2,13,20
5, 4,29,15,57,54,10,50, 4,37,23,48,15,78,23,42,13,20
5, 4,20,10,25,23, 5,38,47,48,17,44,26,52,48
5, 3,45
5, 3,24,26,40
5, 2,19,51,10, 4,48
5, 2, 8,24,55,58, 5,43, C,27, 9,20,24,38, 9,52,58, 2,43,16, 4, 3,37,17, 2,13,20
5, 2, 5,21,14,27,49,11, 4,51,35, 5,32,76,39,54, 8,26,15
5, 1,59,23,35,48,28,48
5, 1,44,54,39, 3,42,48, 9,50,37,30
5, 1,24,29,26,40
5, 1, 4, 5,37,11, 3,44,25, 1,14, 4,26,40
5, 0,2C,19,28,56,43, 7,30
5
4,59,35,41,53,34,48,53,20
4,58,35,54,14,24
4,58,21,35, 3,10,26,21,18,52,25,46,12,56,57,11,15
4,58,15,41,49,26,24
4,58, 1,23,36,21,26,43, 7,30
4,57,41,13,31,31,21,28,53,20
4,57,21, 4,48,34,37,46, 5,27, 8,42,54,79, 8, 8,53,20
4,57,12,12, 3,13,38,47,43,52,19,12
4,56,37,51, 5,37,30
4,56,17,46,40
4,55,57,43,35,52,54,12,40,29,37,46,40
4,55,14,42
4,54,54,43,12
4,54,4C,34,37,12,31,57,20,51,46,56, 0,56,15
4,54,20,38, 7,45,37,30
4,54, 0,42,59,16,53,48,31,56, 2,57,46,40
4,53,40,49,11,40,52, 7, 0,11,59,43, 7, 8,46,33,57,51,36,17,46,40
4,53,32, 3, 0,57,55,21,12,57,36
4,53,17,58,24, 2,53,21,51,19,41,15
4,52,38,17,56,32,35,33,20
4,52, 9,46, 0,22,10,13,14,41,33,49,52,12,17,16,48
4,51,36
4,51,16,16
4,50,42,36,10,37,39,15,33,20
4,50,22,55,47,26,19, 4,13,45,43,40, 1,38,45,55,33,20
4,50,14,15,31,16,36,28,48
4,50, 3,16,44, 7,46,17,17,14, 4, 9,59,39, 2,17,14,55,24,44,13,29,52,35,33,20
4,49,54,37, 3,10,32,26,52,48
4,49,4C,42,51,53,57,53,26,15
4,49,21, 6,40
4,49, 1,31,47,41,49,11,26,25,11, 6,40
4,48,19,30,42,11,15
4,48
4,47,40,30,37, 2,13,20
4,46,47,49,55, 0, 4, 1,12,51, 5,21, 0,53, 6, 5,58,50,51,51, 6,40
4,46,35,16, 4,13,26,24
4,46,25,31,15, 2,49,18, 3,43, 7,56,22, 1,52,30
4,46, 6, 8,15,42,11,15
4,45,46,46,35, 3,42,13,2C
4,45,27,26,13, 2, 2,39,26,50, 3,33,59,30,22,13,20
4,45,18,54,46,17,54, 2,37,19, 1,37,55,12
4,44,45,56,15
4,44,26,40
4,44, 7,25, 3,14,47,14,34, 4,26,40
4,43,26, 6,43,12
```

P	Q	R	
1	5	13	12,42,56,22, 1,52,30
43	5	0	12,43,32,59, 4, 9,59, 2,24
26	12	0	12,44,24,42,51,15,50,24
9	19	0	12,45,16,30, 8,38,24
0	30	4	12,46, 8,20,56,31,54,36,33,45
1	3	22	12,46,31,14,17,42,33,54,22,30
37	0	6	12,47, 8, 1,38,45,55,33,20
8	1	0	12,48
1	13	5	12,48,52, 1,52,30
10	0	10	12,51,36,17,46,40
58	3	0	12,52,13,19,46,11,56, 3,58, 4,48
1	11	14	12,52,28,34,18,23,54,22,30
41	10	0	12,53, 5,38,48,28, 6,31,40,48
24	17	0	12,53,58, 1,23,24,17,16,48
7	24	0	12,54,50,27,31,14,52,48
14	0	21	12,55,13,36,28,20,24,41,28,53,20
25	0	1	12,56,43,22,40
6	6	0	12,57,36
1	19	6	12,58,28,40,53,54,22,30
29	0	12	13, 0,22, 7,50,46,54,48,53,20
0	1	6	13, 1,15
1	17	15	13, 2, 7,55,44, 7,42,18,16,52,30
39	15	0	13, 2,45,28, 2,34,27,36,34,33,36
22	22	0	13, 3,38,29,54,26,50,29,45,36
5	29	0	13, 4,31,35,21,53,18,57,36
2	0	16	13, 4,55, 1,40,41,40
44	0	3	13, 5,32,42, 0,20, 8,53,20
21	4	0	13, 6,25,55,12
4	11	0	13, 7,19,12
1	25	7	13, 8,12,32,24,34,48,16,52,30
17	0	7	13,10, 7,24,26,40
0	7	7	13,11, 0,56,15
20	27	0	13,13,26,13,46,52,40,37,37,55,12
21	0	18	13,13,49,56, 4, 3,37,17, 2,13,20
3	34	0	13,14,19,59, 3,24,43,56,49,12
0	5	16	13,14,43,42,56,57,11,15
36	2	0	13,15,21,51,31,50,24
19	9	0	13,16,15,44,38,24
2	16	0	13,17, 9,41,24
0	3	25	13,18,27,32,23,26,50,19, 8,26,15
36	0	9	13,19, 5,51,42,52,50,22,13,20
5	0	2	13,20
0	13	8	13,20,54,11,57,11,15
9	0	13	13,23,44,18,48,57,36
1	39	0	13,24,15,44, 2,42,17,29,46,48,54
51	0	0	13,24,23,53, 5,37,25,54, 8
0	11	17	13,24,39,45,44, 9,54, 8,26,15
34	7	0	13,25,18,22,55,29,16,48
17	14	0	13,26,12,56,26,52,48
0	21	0	13,27, 7,33,40, 3
24	0	4	13,29, 5,11, 6,40
1	4	1	13,30
0	19	9	13,30,54,52,36, 9, 8,26,15
28	0	15	13,32,53, 3,10,23,52, 5,55,33,20
1	2	10	13,33,48, 7,30
49	5	0	13,34,27,11, 0,26,38,58,33,36
32	12	0	13,35,22,21,42,40,53,45,36
15	19	0	13,36,17,36, 9,12,57,36
0	27	1	13,37,12,54,20,18, 2,15
1	0	19	13,37,37,19,14,53,24,10
43	0	6	13,38,16,33,45,20,59,15,33,20
14	1	0	13,39,12
1	10	2	13,40, 7,30
0	25	10	13,41, 3, 3,45,36,15,17,34,41,15
16	0	10	13,43, 2,42,57,46,40
1	8	11	13,43,58,28,35,37,30
47	10	0	13,44,38, 1,23,41,58,57,47,31,12
30	17	0	13,45,33,53,28,57,54,25,55,12
13	24	0	13,46,29,49,21,19,52,19,12
0	33	2	13,47,25,49, 1, 3,15,46,41,15
1	6	20	13,47,50,32,14,19,34,13, 7,30
31	0	1	13,48,30,16,10,40
12	6	0	13,49,26,24
1	16	3	13,50,22,35,37,30
35	0	12	13,52,23,36,22,10, 2,28, 8,53,20
4	0	5	13,53,20
1	14	12	13,54,16,27,27, 4,13, 7,30
28	22	0	13,55,53, 3,54, 4,37,51,44,38,24
11	29	0	13,56,49,41,43,20,52,13,26,24
8	0	16	13,57,14,41,47,24,26,40
50	0	3	13,57,54,52,48,21,29,28,53,20
27	4	0	13,58,51,38,52,48
10	11	0	13,59,48,28,48
1	22	4	14, 0,45,22,34,13, 7,30
23	0	7	14, 2,47,54, 4,26,40
0	4	4	14, 3,45
1	20	13	14, 4,42, 9,47,39,31,17,20,37,30
9	34	0	14, 7,17,18,59,38,22,52,36,28,48
0	2	13	14, 7,42,37,48,45
42	2	0	14, 8,23,18,57,57,45,36
25	9	0	14, 9,20,47,36,57,30
8	16	0	14,10,18,20, 9,36
1	28	5	14,11,15,56,36, 8,47,20,37,30
0	0	22	14,11,41,22,33, 0,37,40,25
42	0	9	14,12,22,15, 9,44,21,43,42,13,20
11	0	2	14,13,20
0	10	5	14,14,17,48,45
15	0	13	14,17,20,19,45,11, 6,40
57	0	0	14,18, 1,28,37,59,55,37,44,32
0	8	14	14,18,18,24,47, 6,33,45

4,43, 6,55,52,19,12
4,42,53,21,14, 7,13,52,39,13,42,39,22,30
4,42,34,12,36,15
4,42,15, 5,16 6,37,15,23,27,24,26,40
4,41,55,59,13,36,50, 1,55,23,30,55,47,39,37,30,12,20,44,26,40
4,41,47,34, 5,43,36,20,22, 2,29,45,36
4,41,34, 3,15,53,10,25,46,52,30
4,41,15
4,40,55,58, 1,28,53,20
4,39,56, 9,36
4,39,42,44, 6,43,32,12,28,56,39, 9,34,38,23,36,47,48,45
4,39,37,12,57,36
4,39,23,48,22,50, 6,17,55,46,52,30
4,39, 4,53,55,48, 8,53,20
4,38,46, 0,45,32,27,54,27,36,41,55,13,34,48,53,20
4,38,37,41,18, 1,32,37,14,52,48
4,38, 5,29, 9, 1,24,22,30
4,37,46,40
4,37,27,52, 7,23,20,49,22,57,46,40
4,36,47,31,52,30
4,36,28,48
4,36,10, 5,23,33,20
4,35,56,50,44,46,31,24,22,30
4,35,38,10,18, 4,35,26,44,56,17,46,40
4,35,19,31, 7,12, 3,51,33,56,14,44,10,36,58,39,20,29,37,46,40
4,35,11,17,49,39,18, 8,38,24
4,34,58, 6, 0, 2,42,31,44,22,12,25,18,45
4,34,39,29,31,52,30
4,34,20,54,19,15,33,20
4,34, 2,20,22, 6,45,45, 4, 9,39,25,25,55,33,20
4,33,22,30
4,33, 4
4,32,13,59,48,13,25,22,42,54, 7,11,16,32,35,33,20
4,32, 5,52, 3, 4,19,12
4,31,55,34,26,22,17, 8,42,24,26,24,22,10,20,53,40,14,26,56,27,39,15,33,20
4,31,47,27,14,13,37,55,12
4,31,34,25,11, 9,20,31,20,51,33,45
4,31,16, 2,30
4,30,57,41, 3,27,57,21,58,31, 6,40
4,30,31,15,55,53,51,41, 9, 9,35,46,10,33,36
4,30,18,17,32, 3, 2,48,45
4,30
4,29,41,43,42,13,20
4,28,44,18,48,57,36
4,28,34, 8,49,44,58,24,53,44, 8,18, 8,33,55,27, 4,55,45, 7,36,56,33, 8,28,38,31, 6,/   /40
4,28,31,25,32,51,23,43,10,59,11,11,35,39,15,28, 7,30
4,28,26, 7,38,29,45,36
4,28,13,15,14,43,18, 2,48,45
4,27,55, 6,10,22,13,20
4,27,36,58,19,43, 9,59,28,54,25,50,37, 2,13,20
4,26,58, 3,59, 3,45
4,26,40
4,26,21,57,14,17,36,47,24,26,40
4,25,43,13,48
4,25,25,14,52,48
4,25,12,31, 9,29,16,45,36,46,36,14,24,50,37,30
4,24,54,34,18,59, 3,45
4,24,36,38,41,21,12,25,40,44,26,40
4,24,18,44,16,30,46,54,18,10,47,44,48,25,53,54,34, 4,26,40
4,24,1C,50,42,52, 7,49, 5,39,50,24
4,23,58,10,33,38,36, 1,40,11,43, 7,30
4,23,4C,18,45
4,23,22,28, 8,53,2C
4,23, 4,38,45,13,41,31,15,59,40,14,48,53,20
4,22,26,24
4,22, 8,38,24
4,21,56, 4, 6,24,28,24,18,32,41,43, 7,30
4,21,38,20,33,33,53,20
4,21,20,38,12,41,41, 9,48,23, 9,18, 1,78,53,20
4,21, 2,57, 3,42,59,39,33,30,39,44,59,41, 8, 3,31,25,52,15,48, 8,53,20
4,20,55, 9,20,51,29,12,11,31,12
4,20,42,38,34,42,34, 6, 5,37,30
4,20,25
4,20, 7,22,36,55,38,16,17,46,40
4,19,29,33,37,58, 7,30
4,19,12
4,18,54,27,33,20
4,18,24,32, 9,26,48,13,49,37,46,40
4,18, 7, 2,55,30, 3,33,58,48,54,47,47,29,22,57,46,40
4,17,55,20,27,48, 5,45,36
4,17,46,58, 7,32,32,22,15,20,49, 8,43,49,41,15
4,17,29,31,26, 7,58, 7,30
4,17,12, 5,55,33,20
4,16,54,41,35,43,50,23,30, 9, 3,12,35,33,20
4,16,17,20,37,30
4,16
4,15,52,40,32,55,18,31, 6,40
4,14,55,51, 2,13,23,34,24,45,24,45,20,47,12, 5,18,58,32,45,25,55,33,20
4,14,48,14,17, 5,16,48
4,14,36, 1, 6,42,30,29,23,18,20,23,26,15
4,14,18,47,20,37,30
4,14, 1,34,44,29,57,31,51, 6,40
4,13,44,23,18,15, 9, 1,43,51, 9,50,12,53,39,45,11, 6,40
4,13,36,48,41, 9,14,42,19,50,14,47, 2,24
4,13,24,38,56,17,51,23,12,11,15
4,13, 7,30
4,12,50,22,13,20
4,11,56,32,38,24
4,11,44,27,42, 3,10,59,14, 2,59,14,37,10,33,15, 7, 1,52,30
4,11,35,29,39,50,24

```
 P  Q  R
40  7  0   14,18,59,36,27,11,13,55,12
23 14  0   14,19,57,48,12,40,19,12
 6 21  0   14,20,56, 3,54,43,12
 0  6 23   14,22,20, 8,34,55,23, 8,40,18,45
30  0  4   14,23, 1,31,51, 6,40
 5  3  0   14,24
 0 16  6   14,24,58,32, 6,33,45
 3  0  8   14,28, 3,20
55  5  0   14,28,44,59,44,28,25,34,27,50,24
 0 14 15   14,29, 2, 8,35,41,53,40,18,45
38 12  0   14,29,43,51, 9,31,37,20,38,24
21 19  0   14,30,42,46,33,49,49,26,24
 4 26  0   14,31,41,45,57,39,14,24
 7  0 19   14,32, 7,48,31,52,57,46,40
49  0  6   14,32,49,40, 0,22,23,12,35,33,20
20  1  0   14,33,48,48
 3  8  0   14,34,48
 0 22  7   14,35,47,16, 0,38,40,18,45
22  0 10   14,37,54,53,49,37,46,40
 1  5  8   14,38,54,22,30
36 17  0   14,40,36, 9, 2,53,46, 3,38,52,48
19 24  0   14,41,35,48,38,45,11,48,28,48
 2 31  0   14,42,35,32,17, 7,28,49,48
 1  3 17   14,43, 1,54,23,16,52,30
37  0  1   14,43,44,17,15,22,40
18  6  0   14,44,44, 9,36
 1 13  0   14,45,44, 6
 0 28  8   14,46,44, 6,27,39, 9,18,59, 3,45
 1  1 26   14,47,10,35,59,23, 9,14,36, 2,30
10  0  5   14,48,53,20
 1 11  9   14,49,53,33,16,52,30
17 29  0   14,52,37, 0,30,14,15,42,20, 9,36
14  0 16   14,53, 3,40,34,34, 4,26,40
 0 36  0   14,53,37,28,56,20,19,26,25,21
56  0  3   14,53,46,32,19,34,55,26,48,53,20
 1  9 18   14,54, 4,10,49, 4,20, 9,22,30
33  4  0   14,54,47, 5,28,19,12
16 11  0   14,55,47,42,43,12
 1 19  1   14,56,48,24, 4,30
29  0  7   14,58,59, 5,40,44,26,40
 0  1  1   15
 1 17 10   15, 1, 0,58,26,50, 9,22,30
 2  0 11   15, 4,13,28,20
48  2  0   15, 4,56,52,13,49,36,38,24
31  9  0   15, 5,58,10,47,25,26,24
14 16  0   15, 6,59,33,30,14,24
 1 25  2   15, 8, 1, 0,22,33,22,30
 6  0 22   15, 8,28% 8, 3,12,40,11, 6,40
17  0  2   15,10,13,20
 0  7  2   15,11,15
 1 23 11   15,12,16,44,10,40,16,59,31,52,30
21  0 13   15,14,29,41, 4,11,51, 6,40
63  0  0   15,15,13,34,32,31,55,20,15,30, 8
 0  5 11   15,15,31,38,26,15
46  7  0   15,16,15,34,52,59,58,50,52,48
29 14  0   15,17,17,39,25,31, 0,28,48
12 21  0   15,18,19,48,10,22, 4,48
 1 31  3   15,19,22, 1, 7,50,17,31,52,30
 0  3 20   15,19,49,29, 9,15, 4,41,15
36  0  4   15,20,33,37,58,31, 6,40
11  3  0   15,21,36
 0 13  3   15,22,38,26,15
 9  0  8   15,25,55,33,20
 0 11 12   15,26,58,17,10, 4,41,15
44 12  0   15,27,42,46,34, 9,43,50, 0,57,36
27 19  0   15,28,45,37,40, 5, 8,44, 9,36
10 26  0   15,29,48,33, 1,29,51,21,36
13  0 19   15,30,16,19,46, 0,29,37,46,40
 0  9 21   15,31,19,21,16, 7, 0,59,45,56,15
26  1  0   15,32, 4, 3,12
 9  8  0   15,33, 7,12
 0 19  4   15,34,10,25, 4,41,15
28  0 10   15,36,26,33,24,56,17,46,40
 1  2  5   15,37,30
 0 17 13   15,38,33,30,52,57,14,45,56,15
25 24  0   15,40,22,11,53,20,12,35,42,43,12
 8 31  0   15,41,25,54,26,15,58,45, 7,12
 1  0 14   15,41,54, 2, 0,50
43  0  1   15,42,39,14,24,24,10,40
24  6  0   15,43,43, 6,14,24
 7 13  0   15,44,47, 2,24
 0 25  5   15,45,51, 2,53,29,45,56,15
 5  0 25   15,46,19,18,23,20,41,51,34,26,40
16  0  5   15,48, 8,53,20
 1  8  6   15,49,13, 7,30
20  0 16   15,52,35,55,16,52,20,44,26,40
 6 36  0   15,53,11,58,52, 5,40,44,11, 2,24
 1  6 15   15,53,40,27,32,20,37,30
39  4  0   15,54,26,13,50,12,28,48
22 11  0   15,55,30,53,34, 4,48
 5 18  0   15,56,35,37,40,48
 0 31  6   15,57,40,26,10,39,53,15,42,11,15
 1  4 24   15,58, 9, 2,52, 8,12,22,58, 7,30
35  0  7   15,58,55, 2, 3,27,24,26,40
 4  0  0   16
 1 14  7   16, 1, 5, 2,20,37,30
 8  0 11   16, 4,30,22,13,20
54  2  0   16, 5,16,39,42,44,55, 4,57,36
 1 12 16   16, 5,35,42,52,59,52,58, 7,30
37  9  0   16, 6,22, 3,30,35, 8, 9,36
```

```
4,11,27,25,32,33, 5,40, 8,12,11,15
4,11,1C,24,32,13,20
4,10,53,24,40,59,13, 7, 0,51, 1,43,42,13,20
4,10,28,56,58,25,25,38, 6,15,33, 7,12
4,10,16,56,14, 7,15,56,15
4,10
4, 9,43, 4,54,39, C,44,26,40
4, 8,45,55,12
4, 8,37,59,12,38,41,57,45,43,41,28,30,47,27,39,22,30
4, 8,33, 4,51,12
4, 8,21, 9,40,17,52,15,56,15
4, 8, 4,21,16,16, 7,54, 4,26,40
4, 7,47,34, 0,28,51,28,24,32,37,15,45,24,16,47,24,26,40
4, 7,40,10, 2,41,22,19,46,33,36
4, 7,28,17,24, 2,26,16,33,55,59,10,46,52,30
4, 7,11,32,34,41,15
4, 6,54,48,53,20
4, 6,38, 6,19,54, 5,10,33,44,41,28,53,?0
4, 6, 2,15
4, 5,45,36
4, 5,17,11,46,28, 1,15
4, 5, 0,35,49,24, 4,50,26,36,42,28, 8,53,20
4, 4,44, 0,59,44, 3,25,50, 9,59,45,55,57,18,48,18,13, 0,14,48,53,20
4, 4,3C,42,3C,48,16, 7,40,48
4, 4,24,58,40, 2,24,28,12,46,24,22,30
4, 4, 8,26,15
4, 3,51,54,57, 7, 9,37,46,40
4, 3,35,24,46,19,2C,40, 3,41,55, 2,36,22,42,57,46,40
4, 3,28, 8,20,18,28,31, 2,14,38,11,33,30,14,24
4, 3
4, 2,43,33,20
4, 1,59, 6,29,31,55,53,31,28, 6,23,21,?2,18,16,17,46,40
4, 1,51,52,56, 3,50,24
4, 1,42,43,56,46,28,34,24,21,43,28,19,42,31,54,22,26,10,36,51,14,53,49,37,46,40
4, 1,40,16,59,34,15,20,51,53,16, 4,26, 5,19,55,18,45
4, 1,35,30,52,38,47, 2,24
4, 1,23,55,43,14,58,14,31,52,30
4, 1, 7,35,33,20
4, 0,51,16,29,44,50,59,32, 0,59,15,33,?0
4, 0,16,15,35, 9,22,30
4
3,59,43,45,30,51,51, 6,40
3,58,52,43,23,31,12
3,58,41,16, 2,32,21, 5, 3, 5,56,36,58,21,33,45
3,58,25, 6,53, 5, 9,22,30
3,58, 8,58,49,13, 5,11, 6,40
3,57,52,51,50,51,42,12,52,21,42,58,19,35,18,31, 6,40
3,57,45,45,38,34,55, 2,11, 5,51,21,36
3,57,18,16,52,30
3,57, 2,13,20
3,56,46,10,52,42,19,22, 8,23,42,13,20
3,56,11,45,36
3,56, 0,25,58,10,29, 3, 1,55,18, 2,27,?1, 8,40,25,20,30,28, 7,30
3,55,56,46,33,36
3,55,44,27,41,46, 1,33,52,41,25,32,48,45
3,55,28,30,30,12,30
3,55,12,34,23,25,31, 2,49,32,50,22,13,20
3,54,56,39,21,20,41,41,36, 9,35,46,29,43, 1,15,10,17,17, 2,13,20
3,54,45,38,24,46,20,16,58,22, 4,48
3,54,38,22,43,14,18,41,29, 3,45
3,54,22,30
3,54, 6,38,21,14, 4,26,40
3,53,16,48
3,53, 1, 0,48
3,52,45,50,19, 1,45,14,56,29, 3,45
3,52,34, 4,56,30, 7,24,26,40
3,52,18,20,37,57, 3,15,23, 0,34,56, 1,19, 0,44,26,40
3,52,11,24,25, 1,17,11, 2,24
3,51,55,41,38,32,25,57,30,14,24
3,51,44,34,17,31,10,18,45
3,51,28,53,20
3,51,13,13,26, 9,27,21, 9, 8, 8,53,20
3,50,39,36,33,45
3,50,24
3,50, 8,24,29,37,46,40
3,49,41,48,35, 3,49,32,17,26,54,48,53,20
3,49,26,15,56, 0, 3,12,58,16,52,16,48,42,28,52,47, 4,41,28,53,20
3,49,19,24,51,22,45, 7,12
3,49, 8,25, 0, 2,15,26,26,58,30,21, 5,?7,30
3,48,52,54,36,33,45
3,48,37,25,16, 2,57,46,40
3,48,21,56,58,25,38, 7,33,28, 2,51,11,36,17,46,40
3,47,59, 7,49, 2,19,14, 5,51,13,18,20, 9,36
3,47,48,45
3,47,33,20
3,46,44,53,22,33,36
3,46,36,18,41,58,34,17,15,20,22, 0,18,?8,37,24,43,32, 2,27, 3, 2,42,57,46,40
3,46,2S,32,41,51,21,36
3,46,18,40,59,17,47, 6, 7,22,58, 7,30
3,46, 3,22, 5
3,45,48, 4,12,53,17,48,18,45,55,33,20
3,45,32,47,22,53,28, 1,32,18,48,44,38, 7,42, 0, 9,52,35,33,20
3,45,26, 3,16,34,53, 4,17,37,59,48,28,48
3,45,15,14,36,42,32,20,37,30
3,45
3,44,44,46,25,11, 6,40
3,43,56,55,40,48
3,43,46,11,17,22,49,45,59, 9,19,19,39,42,42,53,26,15
3,43,41,46,22, 4,48
3,43,31, 2,42,16, 5, 2,20,37,30
```

P	Q	R		
20	16	0	16, 7,27,31,44,15,21,36	3,43,15,55, 8,38,31, 6,40.
3	23	0	16, 8,33, 4,24, 3,36	3,43, C,48,36,25,58,19,34, 5,21,32,10,51,51, 6,40
12	0	22	16, 9, 2, 0,35,25,30,51,51, 6,40	3,42,54, 9, 2,25,14, 5,47,54,14,24
23	0	2	16,10,54,13,20	3,42,28,23,19,13, 7,30
2	5	0	16,12	3,42,13,20
1	20	8	16,13, 5,51, 7,22,58, 7,30	3,41,58,17,41,54,40,39,3C,22,13,20
27	0	13	16,15,27,39,48,28,38,31, 6,40	3,41,26, 1,30
0	2	8	16,16,33,45	3,41,11, 2,24
52	7	0	16,17,20,37,12,31,58,46,16,19,12	3,41, C,25,57,54,23,58, 0,38,50,12, 0,42,11,15
35	14	0	16,18,26,50, 3,13, 4,30,43,12	3,40,45,28,35,49,13, 7,30
18	21	0	16,19,33, 7,23, 3,33, 7,12	3,40,3C,32,14,27,40,21,23,57, 2,13,20
1	28	0	16,20,39,29,12,21,38,42	3,40,15,36,53,45,39, 5,15, 8,59,47,20,?1,34,55,28,23,42,13,20
0	0	17	16,21, 8,47, 5,52, 5	3,40, 5, 2,15,43,26,30,54,43,12
42	0	4	16,21,55,52,30,25,11, 6,40	3,39,58,28,48, 2,1C, 1,23,29,45,56,15
17	3	0	16,23, 2,24	3,39,43,35,37,30
0	10	0	16,24, 9	3,39,28,43,27,24,26,40
1	26	9	16,25,15,40,30,43,30,21, 5,37,3C	3,39,13,52,17,41,24,36, 3,19,43,32,20,44,26,40
15	0	8	16,27,39,15,33,20	3,38,42
0	8	9	16,28,46,10,18,45	3,38,27,12
33	19	0	16,30,40,40,10,45,29,19, 6,14,24	3,38, 1,57, 7,58,14,26,40
16	26	0	16,31,47,47,13,35,50,47, 2,24	3,37,47,11,5C,34,44,18,10,19,17,45, 1,?4, 4,26,40
19	0	19	16,32,17,25, 5, 4,31,36,17,46,40	3,37,4C,41,38,27,27,21,36
1	34	1	16,32,54,58,49,15,54,56, 1,30	3,37,32,27,33, 5,49,42,57,55,33, 7,29,44,16,42,56,11,33,33,10, 7,24,26,40
0	6	18	16,33,24,38,41,11,29, 3,45	3,37,25,57,47,22,54,20, 9,36
32	1	0	16,34,12,19,24,48	3,37,15,32, 8,55,28,25, 4,41,15
15	8	0	16,35,19,40,48	3,37, C,50
0	16	1	16,36,27, 6,45	3,26,4C, 8,5C,46,21,53,34,48,53,20
34	0	10	16,38,52,19,38,36, 2,57,46,40	3,36,14,38, 1,38,26,15
3	0	3	16,40	3,36
0	14	10	16,41, 7,44,56,29, 3,45	3,35,45,22,57,46,4C
14	31	0	16,44,11,38, 4, 1, 2,40, 7,40,48	3,35, 5,52,26,15, 3, 0,54,38,19, 0,45,39,49,34,29, 8, 8,53,20
7	0	14	16,44,41,38, 8,53,20	3,34,59,27, 3,10, 4,48
49	0	1	16,45,29,51,22, 1,47,22,40	3,34,49, 8,26,17, 6,58,32,47,20,57,16,?1,24,22,30
0	12	19	16,45,49,42,10,12,22,40,32,48,45	3,34,44,54, 6,47,48,28,48
30	6	0	16,46,37,58,39,21,36	3,34,34,36,11,46,38,26,15
13	13	0	16,47,46,10,33,36	3,34,2C, 4,56,17,46,40
0	22	2	16,48,54,27, 5, 3,45	3,34, 5,34,39,46,31,59,35, 7,32,40,29,37,46,40
22	0	5	16,51,21,28,53,20	3,33,34,27,11,15
1	5	3	16,52,30	3,33,2C
0	20	11	16,53,38,35,45,11,25,32,48,45	3,33, 5,33,47,26, 5,25,55,33,20
26	0	16	16,56, 6,18,57,59,50, 7,24,26,40	3,32,34,35, 2,24
1	3	12	16,57,15, 9,22,30	3,32,2C,11,54,14,24
45	4	0	16,58, 3,58,45,33,18,43,12	3,32,10, 0,55,35,25,24,29,25,16,59,31,52,30
28	11	0	16,59,12,57, 8,21, 7,12	3,31,55,39,27,11,15
11	18	0	17, 0,22, 0,11,31,12	3,31,41,18,57, 4,57,56,32,35,33,20
0	28	3	17, 1,31, 7,55,22,32,48,45	3,31,26,59,25,12,37,31,26,32,38,11,50,44,43, 7,39,15,33,20
1	1	21	17, 2, 1,39, 3,36,45,12,30	3,31,20,40,34,17,42,15,16,31,52,19,12
41	C	7	17, 2,50,42,11,41,14, 4,26,40	3,31,1C,32,26,54,52,49,2C, 9,22,30
10	0	0	17, 4	3,30,56,15
1	11	4	17, 5, 9,22,30	3,30,41,58,31, 6,40
14	0	11	17, 8,48,23,42,13,20	3,29,57, 7,12
60	2	0	17, 9,37,46,21,35,54,45,17,26,24	3,29,47, 3, 5, 2,39, 9,21,42,29,22,10,?8,47,42,35,51,33,45
1	9	13	17, 9,58, 5,44,31,52,30	3,29,42,54,43,12
43	9	0	17,10,47,31,44,37,28,42,14,24	3,29,32,51,17, 7,34,43,26,50, 9,22,30
26	16	0	17,11,57,21,51,12,23, 2,24	3,29,18,40,26,51, 6,40
9	23	0	17,13, 7,16,41,39,50,24	3,29, 4,30,34, 9,2C,55,50,42,31,26,25,'1, 6,40
0	34	4	17,14,17,16,16,19, 4,43,21,33,45	3,28,5C,21,38,58,23,43,38,48,31,47,59,44,54,26,49, 8,41,48,38,31, 6,40
1	7	22	17,14,48,10,17,54,27,46,24,22,30	3,28,44, 7,28,41,11,21,45,12,57,36
29	0	2	17,15,37,50,13,20	3,28,34, 6,51,46, 3,16,52,30
8	5	0	17,16,48	3,28,2C
1	17	5	17,17,58,14,31,52,30	3,28, 5,54, 5,32,30,37, 2,13,20
33	0	13	17,20,29,30,27,42,33, 5,11, 6,40	3,27,35,38,54,22,30
2	0	6	17,21,40	3,27,21,36
1	15	14	17,22,50,34,18,50,16,24,22,30	3,27, 7,34, 2,40
41	14	0	17,23,40,37,23,25,56,48,46, 4,48	3,26,57,38, 3,34,53,33,16,52,30
24	21	0	17,24,51,19,52,35,47,19,40,48	3,26,43,37,43,33,26,35, 3,42,13,20
7	28	0	17,26, 2, 7, 9,11, 5,16,48	3,26,29,38,2C,24, 2,53,40,27,11, 3, 7,?0,13,59,30,22,13,20
6	0	17	17,26,33,22,14,15,33,20	3,26,23,28,22,14,28,36,28,48
48	0	4	17,27,23,36, 0,26,51,51, 6,40	3,26,13,34,30, 2, 1,53,48,16,39,18,59, 3,45
23	3	0	17,28,34,33,36	3,25,59,37, 8,54,22,30
6	10	0	17,29,45,36	3,25,45,40,44,26,40
1	23	6	17,30,56,43,12,46,24,22,30	3,25,31,45,16,35, 4,18,48, 7,14,34, 4,26,40
21	0	8	17,33,29,52,35,33,20	3,25, 1,52,30
0	5	6	17,34,41,15	3,24,48
22	26	0	17,37,34,58,22,30,14,10,10,33,36	3,24,1C,29,51,10, 4, 2, 2,10,35,23,27,24,26,40
5	33	0	17,39, 6,38,44,32,58,35,45,36	3,23,5e,40,49,46,42,51,31,48,19,48,16,37,45,40,15,10,50,12,20,44,26,40
0	3	15	17,39,38,17,15,56,15	3,23,5C,35,25,40,13,26,24
38	1	0	17,40,29, 8,42,27,12	3,23,4C,48,53,22, 0,23,30,38,40,18,45
21	8	0	17,41,40,59,31,12	3,23,27, 1,52,30
4	15	0	17,42,52,55,12	3,23,13,15,47,35,58, 1,28,53,20
1	29	7	17,44, 4,55,45,10,59,10,46,52,30	3,22,59,30,38,56, 7,13,23, 4,55,52,10,?8,55,48, 8,53,20
0	1	24	17,44,36,43,11,15,47, 5,31,15	3,22,53,26,56,55,23,45,51,52,11,49,37,55,12
40	0	10	17,45,27,48,57,10,27, 9,37,46,40	3,22,43,43, 9, 2,17, 6,33,45
9	0	3	17,46,40	3,22,3C
0	11	7	17,47,52,15,56,15	3,22,1C,16,17,46,40
13	0	14	17,51,40,24,41,28,53,20	3,21,33,14, 6,43,12
3	38	0	17,52,20,58,43,36,23,19,42,25,12	3,21,25,36,37,18,43,48,40,18, 6,13,36,25,26,35,18,41,48,50,42,42,24,51,21,28,53,20
55	0	1	17,52,31,50,47,29,54,32,10,40	3,21,23,34, 9,38,32,47,23,14,23,23,41,44,26,36, 5,37,30
0	9	16	17,52,53, 0,58,53,12,11,15	3,21,19,35,43,52,19,12
36	6	0	17,53,44,30,33,59, 2,24	3,21, 9,56,26, 2,28,32, 6,33,45
19	13	0	17,54,57,15,15,50,24	3,20,56,19,37,46,40
2	20	0	17,56,10, 4,53,24	3,20,42,43,44,47,22,29,36,40,49,22,57,46,40
28	0	5	17,58,46,54,48,53,20	3,20,13,32,59,17,48,45
1	2	0	18	3,20
0	17	8	18, 1,13,10, 8,12,11,15	3,19,46,27,55,43,12,35,33,20
1	0	9	18, 5, 4,10	3,19, 3,56, 9,36
51	4	0	18, 5,56,14,40,35,31,58, 4,48	3,18,54,23,22, 6,57,34,12,34,57,10,48,37,58, 7,30
0	15	17	18, 6,17,40,44,37,22, 5,23,26,15	3,18,5C,27,52,57,36
34	11	0	18, 7, 9,48,56,54,31,40,48	3,18,40,55,44,14,17,48,45
17	18	0	18, 8,23,28,12,17,16,48	3,18,27,29, 1, C,54,19,15,33,20

P	Q	R	
0	25	0	18, 9,37,12,27, 4, 3
5	0	20	18,10, 9,45,39,51,12,13,20
47	C	7	18,11, 2, 5, 0,27,59, 0,44,26,4C
16	0	0	18,12,16
1	8	1	18,13,30
C	23	9	18,14,44, 5, 0,48,20,23,26,15
20	0	11	18,17,23,37,17, 2,13,20
1	6	10	18,18,37,58, 7,30
49	9	0	18,19,30,41,51,35,58,37, 3,21,36
32	16	0	18,20,45,11,18,37,12,34,33,36
15	23	0	18,21,59,45,48,26,29,45,36
0	31	1	18,23,14,25,21,24,21, 2,15
1	4	19	18,23,47,22,59, 6, 5,37,30
35	0	2	18,24,40,21,34,13,20
14	5	0	18,25,55,12
1	14	2	18,27,10, 7,30
8	0	6	18,31, 6,40
1	12	11	18,32,21,56,36, 5,37,30
30	21	0	18,34,30,45,12, 6,10,28,59,31,12
13	28	0	18,35,46,15,37,47,49,37,55,12
12	0	17	18,36,19,35,43,12,35,33,20
0	37	2	18,37, 1,51,10,25,24,18, 1,41,15
54	0	4	18,37,13,10,24,28,39,18,31, 6,40
1	10	20	18,37,35,13,31,20,25,11,43, 7,3C
29	3	0	18,38,28,51,50,24
12	10	0	18,39,44,38,24
1	20	3	18,41, 0,30, 5,37,30
27	0	8	18,43,43,52, 5,55,33,20
0	23	3	18,45
1	18	12	18,46,16,13, 3,32,41,43, 7,30
11	33	0	18,49,43, 5,19,31,10,30, 8,38,24
0	0	12	18,50,16,50,25
44	1	0	18,51,11, 5,17,17, 0,48
27	8	0	18,52,27,43,29,16,48
1C	15	0	18,53,44,26,52,48
1	26	4	18,55, 1,15,28,11,43, 7,30
4	0	23	18,55,35,10, 4, 0,50,13,53,20
15	0	3	18,57,46,40
0	8	4	18,59, 3,45
19	0	14	19, 3, 7, 6,20,14,48,53,20
61	0	1	19, 4, 1,58,10,39,54,10,19,22,40
0	6	13	19, 4,24,33, 2,48,45
42	6	0	19, 5,19,28,36,14,58,33,36
25	13	0	19, 6,37, 4,16,53,45,36
8	20	0	19, 7,54,45,12,57,36
1	32	5	19, 9,12,31,24,47,51,54,50,37,3C
C	4	22	19, 9,46,51,26,33,50,51,33,45
34	0	5	19,10,42, 2,28, 8,53,20
7	2	0	19,12
0	14	5	19,13,18, 2,48,45
7	0	9	19,17,24,26,40
57	4	0	19,18,19,59,39,17,54, 5,57, 7,12
0	12	14	19,18,42,51,27,35,51,33,45
40	11	0	19,19,38,28,12,42, 9,47,31,12
23	18	0	19,20,57, 2, 5, 6,25,55,12
6	25	0	19,22,15,41,16,52,19,12
11	0	20	19,22,50,24,42,30,37, 2,13,20
22	0	0	19,25, 5, 4
5	7	0	19,26,24
0	20	6	19,27,43, 1,20,51,33,45
26	0	11	19,30,33,11,46,10,22,13,20
1	3	7	19,31,52,30
0	18	15	19,33,11,53,36,11,33,27,25,18,45
38	16	0	19,34, 8,12, 3,51,41,24,51,50,24
21	23	0	19,35,27,44,51,40,15,44,38,24
4	30	0	19,36,47,23, 2,49,58,26,24
1	1	16	19,37,22,32,31, 2,30
41	0	2	19,38,19, 3, 0,30,13,20
20	5	0	19,39,38,52,48
3	12	0	19,40,58,48
0	26	7	19,42,18,48,36,52,12,25,18,45
3	0	26	19,42,54, 7,59,10,52,19,28, 3,20
14	0	6	19,45,11, 6,40
1	9	8	19,46,31,24,22,30
19	28	0	19,50, 9,20,40,19, 0,56,26,52,48
18	0	17	19,50,44,54, 6, 5,25,55,33,20
2	35	0	19,51,29,58,35, 7, 5,55,13,48
1	7	17	19,52, 5,34,25,25,46,52,30
35	3	0	19,53, 2,47,17,45,36
18	10	0	19,54,23,36,57,36
1	17	0	19,55,44,32, 6
33	0	8	19,58,38,47,34,19,15,33,20
2	0	1	20
1	15	9	20, 1,21,17,55,46,52,30
6	0	12	20, 5,37,57,46,40
0	40	0	20, 6,23,36, 4, 3,26,14,40,13,21
50	1	0	20, 6,35,49,38,26, 8,51,12
1	13	18	20, 6,59,38,36,14,51,12,39,22,3C
33	8	0	20, 7,57,34,23,13,55,12
16	15	0	20, 9,19,24,40,19,12
1	23	1	20,10,41,20,30, 4,30
10	0	23	20,11,17,30,44,16,53,34,48,53,20
21	0	3	20,13,37,46,40
0	5	1	20,15
1	21	10	20,16,22,18,54,13,42,39,22,30
25	0	14	20,19,19,34,45,35,48, 8,53,20
0	3	10	20,20,42,11,15
48	6	0	20,21,40,46,30,39,58,27,50,24
31	13	0	20,23, 3,32,34, 1,20,38,24
14	20	0	20,24,26,24,13,49,26,24

3,18,14, 3,12,23, 5,10,43,38, 5,48,36,19,25,25,55,33,20
3,18, 8, 8, 2, 9, 5,51,49,14,52,48
3,17,58,37,55,13,57, 1,15, 8,47,20,37,30
3,17,45,14, 3,45
3,17,31,51, 6,40
3,17,18,29, 3,55,16, 8,26,59,45,11, 6,40
3,16,49,48
3,16,36,28,48
3,16,27, 3, 4,48,21,18,13,54,31,17,20,37,30
3,16,13,45,25,10,25
3,16, C,28,39,31,15,52,21,17,21,58,31, 6,40
3,15,47,12,47,47,14,44,40, 7,59,48,44,45,51, 2,38,34,24,11,51, 6,40
3,15,41,22, C,38,36,54, 8,38,24
3,15,31,58,56, 1,55,34,34,13, 7,30
3,15,16,45
3,15, 5,31,57,41,43,42,13,20
3,14,24
3,14,1C,50,40
3,13,48,24, 7, 5, 6,10,22,13,20
3,13,35,17,11,37,32,42,49,10,29, 6,41, 5,50,37, 2,13,20
3,13,25,30,20,51, 4,19,12
3,13,22,11, 9,25,10,51,31,29,22,46,39,46, 1,31,29,56,56,29,28,59,55, 3,42,13,20
3,13,2C,13,35,39,24,16,41,30,36,51,32,*2,15,56,15
3,13,16,24,42, 7, 1,37,55,12
3,12,54, 4,26,40
3,12,41, 1,11,47,52,47,37,36,47,24,26,40
3,12,13, 0,28, 7,3C
3,12
3,11,47, 0,24,41,28,53,2C
3,11,11,53,16,4C, 2,40,48,34, 3,34, 0,35,24, 3,59,13,54,34, 4,26,40
3,11, 6,10,42,48,57,36
3,10,57, 0,50, 1,52,52, 2,28,45,17,34,41,15
3,10,44, 5,30,28, 7,30
3,10,31,11, 3,22,28, 8,53,20
3,10,18,17,28,41,21,46,17,53,22,22,39,40,14,48,53,20
3,10,12,36,30,51,56, 1,44,52,41, 5,16,48
3, 9,5C,37,30
3, 9,37,46,40
3, 8,57,24,28,48
3, 8,48,20,46,32,23,14,25,32,14,25,57,52,54,56,20,16,24,22,30
3, 8,44,37,14,52,48
3, 8,35,34, 9,24,49,15, 6, 9, 8,26,15
3, 8,22,48,24,10
3, 8,10, 3,3C,44,24,50,15,38,16,17,46,40
3, 7,57,19,29, 4,33,21,16,55,40,37,11,46,25, 0, 8,13,49,37,46,40
3, 7,51,42,43,49, 4,13,34,41,39,50,24
3, 7,42,42,10,35,26,57,11,15
3, 7,3C
3, 7,17,18,40,59,15,33,2C
3, 6,37,26,24
3, 6,28,29,24,29, 1,28,19,17,46, 6,23, 5,35,44,31,52,30
3, 6,24,48,38,24
3, 6,15,52,15,13,24,11,57,11,15
3, 6, 3,15,57,12, 5,55,33,2C
3, 5,5C,40,30,21,38,36,18,24,27,56,49, 3,12,35,33,20
3, 5,45, 7,32, 1, 1,44,49,55,12
3, 5,23,39,26, 0,56,15
3, 5,11, 6,40
3, 4,5E,34,44,55,33,52,55,18,31, 6,40
3, 4,31,41,15
3, 4,19,12
3, 4, 6,43,35,42,13,20
3, 3,57,53,49,51, C,56,15
3, 3,45,26,52, 3, 3,37,49,57,31,51, 6,40
3, 3,33, 0,44,48, 2,34,22,37,29,49,26,57,59, 6,13,39,45,11, 6,40
3, 3,27,31,53, 6,12, 5,45,36
3, 3,16,44, 0, 1,48,21, 9,34,48,16,52,30
3, 3, 6,19,41,15
3, 2,53,56,12,50,22,13,20
3, 2,41,33,34,44,30,30, 2,46,26,16,57,17, 2,13,20
3, 2,36, 6,15,13,51,23,16,40,58,38,40, 7,40,48
3, 2,15
3, 2, 2,40
3, 1,29,19,52, 8,56,55, 8,36, 4,47,31, 1,43,42,13,20
3, 1,23,54,42, 2,52,48
3, 1,17, 2,57,34,51,25,48,16,17,36,14,46,53,55,46,49,37,57,38,26,10,22,13,20
3, 1,11,38, 9,29, 5,16,48
3, 1, 2,56,47,26,13,40,53,54,22,30
3, 0,5C,41,40
3, 0,38,27,22,18,38,14,39, 0,44,26,40
3, 0,12,11,41,22, 1,52,3C
3
2,59,47,49, 8, 8,53,20
2,59, 9,32,32,38,24
2,59, 2,45,53, 9,58,56,35,49,25,32, 5,42,36,58, 3,17,10, 5, 4,37,42, 5,39, 5,40,44,/
              /26,40
2,59, 0,57, 1,54,15,48,47,19,27,27,43,46,10,18,45
2,58,57,25, 5,39,50,24
2,58,48,50, 9,48,52, 1,52,30
2,58,36,44, 6,54,48,53,20
2,58,19,19,13,56,11,16,38,19,23,31,12
2,57,46,40
2,57,34,38, 9,31,44,31,36,17,46,40
2,57, 8,49,12
2,56,56,49,55,12
2,56,48,20,46,19,31,10,24,31, 4, 9,36,33,45
2,56,3C,22,52,39,22,30
2,56,24,25,47,34, 8,17, 7, 9,37,46,40

P	Q	R	Value
1	29	2	20,25,49,21,30,27, 3,22,30
0	1	19	20,26,25,58,52,20, 6,15
40	0	5	20,27,24,50,38, 1,28,53,20
13	2	0	20,28,48
0	11	2	20,30,11,15
13	0	9	20,34,34, 4,26,40
0	9	11	20,35,57,42,53,26,15
46	11	0	20,36,57, 2, 5,32,58,26,41,16,48
29	18	0	20,38,20,50,13,26,51,38,52,48
12	25	0	20,39,44,44, 1,59,48,28,48
17	0	20	20,40,21,46,21,20,39,30,22,13,20
1	35	3	20,41, 8,43,31,34,53,40, 1,52,30
0	7	20	20,41,45,48,21,29,21,19,41,15
28	0	0	20,42,45,24,16
11	7	0	20,44, 9,36
0	17	3	20,45,33,53,26,15
32	0	11	20,48,35,24,33,15, 3,42,13,20
1	0	4	20,50
0	15	12	20,51,24,41,10,36,19,41,15
27	23	0	20,53,49,35,51, 6,56,47,36,57,36
10	30	0	20,55,14,32,35, 1,18,20, 9,36
5	0	15	20,55,52, 2,41, 6,40
47	0	2	20,56,52,19,12,32,14,13,20
26	5	0	20,58,17,28,19,12
9	12	0	20,59,42,43,12
0	23	4	21, 1, 8, 3,51,19,41,15
20	0	6	21, 4,11,51, 6,40
1	6	5	21, 5,37,30
0	21	13	21, 7, 3,14,41,29,16,56, 0,56,15
24	0	17	21,10, 7,53,42,29,47,39,15,33,20
8	35	0	21,10,55,58,29,27,34,18,54,43,12
1	4	14	21,11,33,56,43, 7,30
41	3	0	21,12,34,58,26,56,38,24
24	10	0	21,14, 1,11,25,26,24
7	17	0	21,15,27,30,14,24
0	29	5	21,16,53,54,54,13,11, 0,56,15
1	2	23	21,17,32, 3,49,30,56,30,37,30
39	0	8	21,18,33,22,44,36,32,35,33,20
8	0	1	21,20
1	12	6	21,21,26,43, 7,30
12	0	12	21,26, 0,29,37,46,40
56	1	0	21,27, 2,12,56,59,53,26,36,48
1	10	15	21,27,27,37,10,39,50,37,30
39	8	0	21,28,29,24,40,46,50,52,48
22	15	0	21,29,56,42,19, 0,28,48
5	22	0	21,31,24, 5,52, 4,48
27	0	3	21,34,32,17,46,40
4	4	0	21,36
1	18	7	21,37,27,48, 9,50,37,30
31	0	14	21,40,36,53, 4,38,11,21,28,53,20
0	0	7	21,42, 5
54	6	0	21,43, 7,29,36,42,38,21,41,45,36
1	16	16	21,43,33,12,53,32,50,30,28, 7,30
37	13	0	21,44,35,46,44,17,26, 0,57,36
20	20	0	21,46, 4, 9,50,44,44, 9,36
3	27	0	21,47,32,38,56,28,51,36
4	0	18	21,48,11,42,47,49,26,40
46	0	5	21,49,14,30, 0,33,34,48,53,20
19	2	0	21,50,43,12
2	9	0	21,52,12
1	24	8	21,53,40,54, 0,58, 0,28, 7,30
19	0	9	21,56,52,20,44,26,40
0	6	8	21,58,21,33,45
35	18	0	22, 0,54,13,34,20,39, 5,28,19,12
18	25	0	22, 2,23,42,58, 7,47,42,43,12
1	32	0	22, 3,53,18,25,41,13,14,42
0	4	17	22, 4,32,51,34,55,18,45
34	0	0	22, 5,36,25,53, 4
17	7	0	22, 7, 6,14,24
0	14	0	22, 8,36, 9
0	2	26	22,10,45,53,59, 4,43,51,54, 3,45
38	0	11	22,11,49,46,11,28, 3,57, 2,13,20
7	4	0	22,13,20
0	12	9	22,14,50,19,55,18,45
16	30	0	22,18,55,30,45,21,23,33,30,14,24
11	0	15	22,19,35,30,51,51, 6,40
1	38	1	22,20,20,26,13,24,30,29, 9,38, 1,30
53	0	2	22,20,39,48,29,22,23,10,13,20
0	10	18	22,21, 6,16,13,36,30,14, 3,45
32	5	0	22,22,10,38,12,28,48
15	12	0	22,23,41,34, 4,48
0	20	1	22,25,12,36, 6,45
26	0	6	22,28,28,38,31, 6,40
1	3	2	22,30
0	18	10	22,31,31,27,40,15,14, 3,45
1	1	11	22,36,20,12,30
47	3	0	22,37,25,18,20,44,24,57,36
30	10	0	22,38,57,16,11, 8, 9,36
13	17	0	22,40,29,20,15,21,36
0	26	2	22,42, 1,30,33,50, 3,45
3	0	21	22,42,42,12, 4,49, 0,16,40
45	0	8	22,43,47,36,15,34,58,45,55,33,20
14	0	1	22,45,20
1	9	3	22,46,52,30
0	24	11	22,48,25, 6,16, 0,25,29,17,48,45
18	0	12	22,51,44,31,36,17,46,40
62	1	0	22,52,50,21,48,47,53, 0,23,15,12
1	7	12	22,53,17,27,39,22,30
45	8	0	22,54,23,22,19,29,58,16,19,12
28	15	0	22,55,56,29, 8,16,30,43,12

2,56,12,29,31, 0,31,16,12, 7,11,49,52,17,15,56,22,42,57,46,40
2,56, 7,13,48,34,45,12,43,46,33,36
2,55,58,47, 2,25,44, 1, 6,47,48,45
2,55,46,52,30
2,55,34,58,45,55,33,20
2,54,57,36
2,54,45,45,36
2,54,37,22,44,16,18,56,12,21,47,48,45
2,54,25,33,42,22,35,33,20
2,54,13,45,28,27,47,26,32,15,26,12, 0,59,15,33,20
2,54, 8,33,18,45,57,53,16,48
2,54, 1,58, 2,28,39,46,22,20,26,29,59,47,25,22,20,57,14,50,32, 5,55,33,20
2,53,56,46,13,54,19,28, 7,40,48
2,53,48,25,43, 8,22,44, 3,45
2,53,36,40
2,53,24,55, 4,37, 5,30,51,51, 6,40
2,52,59,42,25,18,45
2,52,48
2,52,36,18,22,13,20
2,52,16,21,26,17,52, 9,13, 5,11, 6,40
2,52, 4,41,57, 0, 2,24,43,42,39,12,36,11,51,39,35,18,31, 6,40
2,51,59,33,38,32, 3,50,24
2,51,51,18,45, 1,41,34,50,13,52,45,49,13, 7,30
2,51,39,40,57,25,18,45
2,51,28, 3,57, 2,13,20
2,51,16,27,43,49,13,35,40, 6, 2, 8,23,42,13,20
2,50,51,33,45
2,50,40
2,50,28,27, 1,56,52,20,44,26,40
2,50, 3,40, 1,55,12
2,49,57,14, 1,28,55,42,56,30,16,30,13,51,28, 3,32,39, 1,50,17,17, 2,13,20
2,49,52, 9,31,23,31,12
2,49,46, 0,44,28,20,19,35,32,13,35,37,30
2,49,32,31,33,45
2,49,21, 3, 9,39,58,21,14, 4,26,40
2,49, 9,35,32,10, 6, 1, 9,14, 6,33,28,35,46,30, 7,24,26,40
2,49, 4,32,27,26, 9,48,13,13,29,51,21,36
2,48,56,25,57,31,54,15,28, 7,30
2,48,45
2,48,33,34,48,53,20
2,47,57,41,45,36
2,47,49,38,28, 2, 7,19,29,21,59,29,44,47, 2,10, 4,41,15
2,47,46,19,46,33,36
2,47,38,17, 1,42, 3,46,45,28, 7,30
2,47,26,56,21,28,53,20
2,47,15,36,27,19,28,44,40,34, 1, 9, 8, 8,53,20
2,46,51,17,29,24,50,37,30
2,46,40
2,46,28,43,16,26, 0,29,37,46,40
2,46, 4,31, 7,30
2,45,53,16,48
2,45,45,19,28,25,47,58,30,29, 7,39, 0,21,38,26,15
2,45,42, 3,14, 8
2,45,34, 6,26,51,54,50,37,30
2,45,22,54,10,50,45,16, 2,57,46,40
2,45,11,42,40,19,14,18,56,21,44,50,30,16,11,11,36,17,46,40
2,45, 6,46,41,47,34,53,11, 2,24
2,44,58,51,36, 1,37,31, 2,37,19,27,11,15
2,44,47,41,43, 7,30
2,44,36,32,35,33,20
2,44,25,24,13,16, 3,27, 2,29,47,39,15,33,20
2,44, 1,30
2,43,50,24
2,43,31,27,50,58,40,50
2,43,20,23,52,56, 3,13,37,44,28,18,45,55,33,20
2,43, 9,20,39,49,22,17,13,26,39,50,37,18,12,32,12, 8,40, 9,52,35,33,20
2,43, 4,28,20,32,10,45, 7,12
2,42,56,39, 6,41,36,18,48,30,56,15
2,42,45,37,30
2,42,34,36,38, 4,46,25,11, 6,40
2,42,18,45,33,32,19, 0,41,29,45,27,42,20, 9,36
2,42,10,58,31,13,49,41,15
2,42
2,41,49, 2,13,20
2,41,19,24,19,41,17,15,40,58,44,15,34,14,52,10,51,51, 6,40
2,41,14,35,17,22,33,36
2,41, 8,29,17,50,59, 2,56,14,28,58,53, 8,21,16,14,57,27, 4,34, 9,55,53, 5,11, 6,40
2,41, 6,51,19,42,50,13,54,35,30,42,57,23,33,16,52,30
2,41, 3,40,35, 5,51,21,36
2,40,55,57, 8,49,58,49,41,15
2,40,45, 3,42,13,20
2,40,34,10,59,49,53,59,41,20,39,30,22,13,20
2,40,10,50,23,26,40
2,40
2,39,49,10,20,34,34, 4,26,40
2,39,15, 8,55,40,48
2,39, 7,30,41,41,34, 3,22, 3,57,44,38,54,22,30
2,38,56,44,35,23,26,15
2,38,45,59,12,48,43,27,24,26,40
2,38,30,30,25,43,16,41,27,23,54,14,24
2,38,12,11,15
2,38, 1,28,53,20
2,37,50,47,15, 8,12,54,45,35,48, 8,53,20
2,37,27,50,24
2,37,20,17,18,46,59,22, 1,16,52, 1,38,14, 5,46,56,53,40,18,45
2,37,17,11, 2,24
2,37, 9,38,27,50,41, 2,35, 7,37, 1,52,30
2,36,59, 0,20, 8,20

P	Q	R		
11	22	0	22,57,29,42,15,33, 7,12	2,36,48,22,55,37, C,41,53, 1,53,34,48,*3,20
0	32	3	22,59, 3, 1,41,45,26,17,48,45	2,36,37,46,14,13,47,47,44, 6,23,50,59,48,40,50, 6,51,31,21,28,53,20
1	5	21	22,59,44,13,43,52,37, 1,52,30	2,36,33, 5,36,30,53,31,18,54,43,12
33	0	3	23, 0,50,26,57,46,40	2,36,25,35, 8,49,32,27,39,22,30
10	4	0	23, 2,24	2,36,15
1	15	4	23, 3,57,39,22,30	2,36, 4,25,34, 9,22,57,46,40
6	0	7	23, 8,53,20	2,35,31,12
1	13	13	23,10,27,25,45, 7, 1,52,30	2,35,20,40,32
43	13	0	23,11,34, 9,51,14,35,45, 1,26,24	2,35,13,13,32,41,10, 9,57,39,22,30
26	20	0	23,13, 8,26,30, 7,43, 6,14,24	2,35, 2,43,17,40, 4,56,17,46,40
9	27	0	23,14,42,49,32,14,47, 2,24	2,34,52,13,45,18, 2,10,15,20,23,17,20,52,40,29,37,46,40
10	0	18	23,15,24,29,39, 0,44,26,40	2,34,47,36,16,46,51,27,21,36
52	0	5	23,16,31,28, 0,35,49, 8, 8,53,20	2,34,4C,10,52,31,31,25,21,12,29,29,14,17,48,45
25	2	0	23,18, 6, 4,48	2,34,29,42,51,40,46,52,30
8	9	0	23,19,40,48	2,34,19,15,33,20
1	21	5	23,21,15,37,37, 1,52,30	2,34, 8,48,57,26,18,14, 6, 5,25,55,33,20
25	0	9	23,24,39,50, 7,24,26,40	2,34,46,24,22,30
0	3	5	23,26,15	2,33,36
1	19	14	23,27,50,16,19,25,52, 8,54,22,30	2,33,25,36,19,45,11, 6,4C
24	25	0	23,30,33,17,50, 0,18,53,34, 4,48	2,33, 7,52,23,22,33, 1,31,37,56,32,35,33,20
7	32	0	23,32, 8,51,39,23,58, 7,40,48	2,32,57,30,37,20, 2, 8,38,51,14,51,12,78,19,15,11,23, 7,39,15,33,20
0	1	14	23,32,51, 3, 1,15	2,32,52,56,34,15,10, 4,48
40	0	0	23,33,58,51,36,36,16	2,32,45,36,40, 1,30,17,37,59, 0,14, 3,45
23	7	0	23,35,34,39,21,36	2,32,35,16,24,22,30
6	14	0	23,37,10,33,36	2,32,24,56,50,41,58,31, 6,40
1	27	6	23,38,46,34,20,14,38,54,22,30	2,32,14,37,58,57, 5,25, 2,18,41,54, 7,44,11,51, 6,40
2	0	24	23,39,28,57,35, 1, 2,47,21,40	2,32,1C, 5,12,41,32,49,23,54, 8,52,13,76,24
13	0	4	23,42,13,20	2,31,52,30
0	9	6	23,43,49,41,15	2,31,42,13,20
17	0	15	23,48,53,52,55,18,31, 6,40	2,31, 5,55,35, 2,24
5	37	0	23,49,47,58,18, 8,31, 6,16,33,36	2,31, 4,12,27,59, 2,51,30,13,34,40,12,19, 4,66,29, 1,21,38, 2, 1,48,38,31, 6,40
59	0	2	23,50, 2,27,43,19,52,42,54,13,20	2,31, 2,40,37,13,54,35,32,25,47,32,46,18,19,57, 4,13, 7,30
0	7	15	23,50,30,41,18,30,56,15	2,30,59,41,47,54,14,24
38	5	0	23,51,39,20,45,18,43,12	2,30,52,27,19,31,51,24, 4,55,18,45
21	12	0	23,53,16,20,21, 7,12	2,30,42,14,43,20
4	19	0	23,54,53,26,31,12	2,30,32, 2,48,35,31,52,12,30,37, 2,13,20
0	5	24	23,57,13,34,18,12,18,34,27,11,15	2,30,17,22,11, 3,15,22,51,45,19,52,19,12
32	0	6	23,58,22,33, 5,11, 6,40	2,30,10, 9,44,28,21,33,45
3	1	0	24	2,30
0	15	7	24, 1,37,33,30,56,15	2,29,49,50,56,47,24,26,40
5	0	10	24, 6,45,33,20	2,29,17,57, 7,12
53	3	0	24, 7,54,59,34, 7,22,37,26,24	2,29,10,47,31,35,13,10,39,26,12,53, 6,78,28,35,37,30
0	13	16	24, 8,23,34,19,29,49,27,11,15	2,29, 7,50,54,43,12
36	10	0	24, 9,33, 5,15,52,42,14,24	2,29, 0,41,48,1C,43,21,33,45
19	17	0	24,11,11,17,36,23, 2,24	2,28,5C,36,45,45,40,44,26,40
2	24	0	24,12,49,36,36, 5,24	2,28,4C,32,11,24,17,18,53, 2,43,34,21,27,14,34, 4,26,40
9	0	21	24,13,33, 0,53, 8,16,17,46,40	2,28,36, 6, 1,36,49,23,51,56, 9,36
20	0	1	24,16,21,20	2,28,18,55,32,48,45
1	6	0	24,18	2,28, 8,53,20
0	21	8	24,19,38,46,41, 4,27,11,15	2,27,58,51,47,56,27, 6,20,14,48,53,20
24	0	12	24,23,11,29,42,42,57,46,40	2,27,37,21
1	4	9	24,24,50,37,30	2,27,27,21,36
51	8	0	24,26, 0,55,48,47,58, 9,24,28,48	2,27,20,17,18,36,15,58,40,25,53,28, 0,78, 7,30
34	15	0	24,27,40,15, 4,49,36,46, 4,48	2,27,1C,19, 3,52,48,45
17	22	0	24,29,19,41, 4,35,19,40,48	2,27, 0,21,29,38,26,54,15,58, 1,28,53,20
0	29	0	24,30,59,13,48,32,28, 3	2,26,50,24,35,50,26, 3,30, 5,59,51,33,34,23,16,58,55,48, 8,53,20
1	2	18	24,31,43,10,38,48, 7,30	2,26,46, 1,30,28,57,40,36,28,48
39	0	3	24,32,53,48,45,37,46,40	2,26,38,59,12, 1,26,40,55,39,50,37,30
16	4	0	24,34,33,36	2,26,29, 3,45
1	12	1	24,36,13,30	2,26,19, 8,58,16,17,46,40
0	27	9	24,37,53,30,46, 5,15,31,38,26,15	2,26, 9,14,51,47,36,24, 2,13, 9, 1,33,49,37,46,40
1	0	27	24,38,37,39,58,58,35,24,20, 4,10	2,26, 4,53, 0,11, 5, 6,37,20,46,54,56, 6, 8,38,24
12	0	7	24,41,28,53,20	2,25,48
1	10	10	24,43, 9,15,28, 7,30	2,25,36, 8
32	20	0	24,46, 1, 0,16, 8,13,58,39,21,36	2,25,21,18, 5,18,49,37,46,40
15	27	0	24,47,41,40,50,23,46,10,33,36	2,25,11,27,53,43, 9,32, 6,52,51,50, 0,49,22,57,46,40
16	0	18	24,48,26, 7,37,36,47,24,26,40	2,25, 7, 7,45,58,18,14,24
0	35	1	24,49,22,28,13,53,52,24, 2,15	2,25, 1,38,22, 3,53, 8,38,37, 2, 4,59,49,31, 8,37,27,42,22, 6,44,56,17,46,40
1	8	19	24,50, 6,58, 1,47,13,35,37,30	2,24,57,18,31,35,16,13,26,24
31	2	0	24,51,18,29, 7,12	2,24,5C,21,25,56,58,56,43, 7,30
14	9	0	24,52,59,31,12	2,24,4C,33,20
1	18	2	24,54,40,40, 7,30	2,24,30,45,53,50,54,35,43,12,35,33,20
31	0	9	24,58,18,29,27,54, 4,26,40	2,24, 9,45,21, 5,37,30
0	0	2	25	2,24
1	16	11	25, 1,41,37,24,43,35,37,30	2,23,50,15,18,31, 6,40
13	32	0	25, 6,17,27, 6, 1,34, 0,11,31,12	2,23,23,54,57,30, 2, 0,36,25,32,40,30,26,33, 2,59,25,25,55,33,20
4	0	13	25, 7, 2,27,13,20	2,23,19,38, 2, 6,43,12
46	0	0	25, 8,14,47, 3, 2,41, 4	2,23,12,45,37,31,24,39, 1,51,33,58,11, 0,56,15
29	7	0	25, 9,56,57,59, 2,24	2,23, 3, 4, 7,51, 5,37,30
12	14	0	25,11,39,15,50,24	2,22,53,23,17,31,51, 6,40
1	24	3	25,13,21,40,37,35,37,30	2,22,43,43, 6,31, 1,19,43,25, 1,46,59,45,11, 6,40
8	0	24	25,14, 6,53,25,21, 6,58,31, 6,40	2,22,39,27,23, 8,57, 1,18,39,30,48,57,36
19	0	4	25,17, 2,13,20	2,22,22,58, 7,30
0	6	3	25,18,45	2,22,13,20
1	22	12	25,20,27,53,37,47, 8,19,13, 7,30	2,22, 3,42,31,37,23,37,17, 2,13,20
23	0	15	25,24, 9,28,26,59,45,11, 6,40	2,21,43, 3,21,36
0	4	12	25,25,52,44, 3,45	2,21,33,27,56, 9,36
44	5	0	25,27, 5,58, 8,19,58, 4,48	2,21,26,40,37, 3,36,56,19,36,51,19,41,15
27	12	0	25,28,49,25,42,31,40,48	2,21,11, 6,18, 7,30
10	19	0	25,30,33, 0,17,16,48	2,21, 7,32,38, 3,18,37,41,43,42,13,20
1	30	4	25,32,16,41,53, 3,49,13, 7,30	2,20,57,59,36,48,25, 0,57,41,45,27,53,49,48,45, 6,10,22,13,20
0	2	21	25,33, 2,28,35,25, 7,48,45	2,20,53,47, 2,51,48,10,11, 1,14,52,48
38	0	6	25,34,16, 3,17,31,51, 6,40	2,20,47, 1,37,56,35,12,53,26,15
9	1	0	25,36	2,20,37,30
0	12	4	25,37,44, 3,45	2,20,27,59, 0,44,26,40
11	0	10	25,43,12,35,33,20	2,19,58, 4,48
59	3	0	25,44,26,39,32,23,52, 7,56, 9,36	2,19,51,22, 3,21,46, 6,14,28,19,34,47,19,11,48,23,54,22,30
0	10	13	25,44,57, 8,36,47,48,45	2,19,48,36,28,48
42	10	0	25,46,11,17,36,56,13, 3,21,36	2,19,41,54,11,25, 3, 8,57,53,26,15
25	17	0	25,47,56, 2,46,48,34,33,36	2,19,32,26,57,54, 4,26,40

```
 P  Q  R
 8 24  0    25,49,40,55, 2,29,45,36                      2,19,23, 0,22,46,13,57,13,48,20,57,36,47,24,26,40
15  0 21    25,50,27,12,56,40,49,22,57,46,4C             2,19,18,50,39, 0,46,18,37,26,24
 0  8 22    25,52,12,15,26,51,41,39,36,23,45             2,19, 9,24,59, 7,27,34,30, 8,38,24
26  0  1    25,53,26,45,20                               2,19, 2,44,34,30,42,11,15
 7  6  0    25,55,12                                     2,18,53,20
 C 18  5    25,56,57,21,47,48,45                         2,18,43,56, 3,41,40,24,41,28,53,20
30  0 12    26, 0,44,15,41,33,49,37,46,40                2,18,23,45,56,15
 1  1  6    26, 2,30                                     2,18, 5, 2,41,46,40
 0 16 14    26, 4,15,51,28,15,24,36,33,45                2,17,58,25,22,23,15,42,11,15
40 15  0    26, 5,30,56, 5, 8,55,13, 9, 7,12             2,17,49, 5, 9, 2,17,43,22,28, 8,53,20
23 22  0    26, 7,16,59,48,53,40,59,31,12                2,17,39,45,33,36, 1,55,46,58, 7,22, 5,13,29,19,40,14,48,53,20
 6 29  0    26, 9, 3,10,43,46,37,55,12                   2,17,35,38,54,49,39, 4,19,12
 3  0 16    26, 9,50, 3,21,23,20                         2,17,29, 3, 0, 1,21,15,52,11, 6,12,39,?2,30
45  0  3    26,11, 5,24, 0,40,17,46,4C                   2,17,19,44,45,56,15
22  4  0    26,12,51,50,24                               2,17,1C,27, 9,37,46,40
 5 11  0    26,14,38,24                                  2,17, 1,10,11, 3,22,52,32, 4,49,42,42,57,46,40
 0 24  6    26,16,25, 4,49, 9,36,33,45                   2,16,41,15
18  0  7    26,20,14,48,53,20                            2,16,32
 1  7  7    26,22, 1,52,30                               2,16, 6,59,54, 6,42,41,21,27, 3,35,38,16,17,46,40
21 27  0    26,26,52,27,33,45,21,15,15,50,24             2,16, 2,56, 1,32, 9,36
22  0 18    26,27,39,52, 8, 7,14,34, 4,26,4C             2,15,57,47,13,11, 8,34,21,12,13,12,11, 5,10,26,50, 7,13,28,13,49,37,46,40
 4 34  0    26,28,39,58, 6,49,27,53,38,24                2,15,53,43,37, 6,48,57,36
 1  5 16    26,29,27,25,53,54,22,30                      2,15,47,12,35,34,40,15,40,25,46,52,30
37  2  0    26,30,43,43, 3,40,48                         2,15,38, 1,15
20  9  0    26,32,31,29,16,48                            2,15,28,50,31,43,58,40,59,15,33,20
 3 16  0    26,34,19,22,48                               2,15,19,40,25,44, 4,48,55,23,17,14,46,?2,37,12, 5,55,33,20
 0 30  7    26,36, 7,23,37,46,28,46,10,18,45             2,15,15,37,57,56,55,50,34,34,47,53, 5,16,48
 1  3 25    26,36,55, 4,46,53,40,38,16,52,30             2,15, 9, 8,46, 1,31,24,22,30
37  0  9    26,38,11,43,25,45,40,44,26,40                2,15
 6  0  2    26,40                                        2,14,5C,51,51, 6,40
 1 13  8    26,41,48,23,54,22,30                         2,14,22, 9,24,28,48
10  0 13    26,47,30,37, 2,13,20                         2,14,17, 4,24,52,29,12,26,52, 4, 9, 4,16,57,43,32,27,52,33,48,28,16,34,14,19,15,33,/
 2 39  0    26,48,31,28, 5,24,34,59,33,37,48             2,14,15,42,46,25,41,51,35,29,35,35,47,49,37,44, 3,45                               /20
52  0  0    26,48,47,46,11,14,51,48,16                   2,14,13, 3,49,14,52,48
 1 11 17    26,49,19,31,28,19,48,16,52,30                2,14, 6,37,37,21,39, 1,24,22,30
35  7  0    26,50,36,45,50,58,33,36                      2,13,57,33, 5,11, 6,40
18 14  0    26,52,25,52,53,45,36                         2,13,48,29, 9,51,34,59,44,27,12,55,18,31, 6,40
 1 21  0    26,54,15, 7,20, 6                            2,13,25, 1,59,31,52,30
25  0  4    26,58,10,22,13,20                            2,13,20
 0  3  0    27                                           2,13,10,58,37, 8,48,23,42,13,20
 1 19  9    27, 1,49,45,12,18,16,52,3C                   2,12,51,36,54
29  0 15    27, 5,46, 6,20,47,44,11,51, 6,40             2,12,42,37,26,24
 0  1  9    27, 7,36,15                                  2,12,36,15,34,44,38,22,48,23,18, 7,12,25,18,45
50  5  0    27, 8,54,22, 0,53,17,57, 7,12                2,12,27,17, 9,29,31,52,3C
33 12  0    27,10,44,43,25,21,47,31,12                   2,12,16,19,20,40,36,12,5C,22,13,20
16 19  0    27,12,35,12,18,25,55,12                      2,12, 9,22, 8,15,23,27, 9, 5,23,52,24,12,56,57,17, 2,13,20
 1 27  1    27,14,25,48,40,36, 4,30                      2,12, 5,25,21,26, 3,54,32,49,55,12
 2  0 19    27,15,14,38,29,46,48,20                      2,11,59, 5,16,49,18, 0,5C, 5,51,33,45
44  0  6    27,16,33, 7,30,41,58,31, 6,40                2,11,5C, 9,22,30
15  1  0    27,18,24                                     2,11,41,14, 4,26,40
 0  9  1    27,20,15                                     2,11,32,19,22,36,50,45,37,59,50, 7,24,26,40
 1 25 10    27,22, 6, 7,31,12,30,35, 9,22,3C             2,11,13,12
17  0 10    27,26, 5,25,55,33,20                         2,11, 4,19,12
 0  7 10    27,27,56,57,11,15                            2,1C,58, 2, 3,12,14,12, 9,16,20,51,33,^5
48 10  0    27,29,16, 2,47,23,57,55,35, 2,24             2,10,45,10,16,46,56,40
31 17  0    27,31, 7,46,57,55,48,51,50,24                2,10,4C,19, 6,20,5C,34,54,11,34,39, 0,44,26,40
14 24  0    27,32,59,38,42,39,44,38,24                   2,10,31,28,31,51,29,49,46,45,19,52,29,50,34, 1,45,42,56, 7,54, 4,26,40
 1 33  2    27,34,51,38, 2, 6,31,33,22,30                2,10,27,34,40,25,44,36, 5,45,36
 0  5 19    27,35,41, 4,28,39, 8,26,15                   2,10,21,19,17,21,17, 3, 2,48,45
32  0  1    27,37, 0,32,21,20                            2,10,12,30
13  6  0    27,38,52,48                                  2,10, 3,41,18,27,49, 8, 8,53,20
 0 15  2    27,40,45,11,15                               2, 9,44,46,48,59, 3,45
36  0 12    27,44,47,12,44,20, 4,56,17,46,40             2, 9,3é
 5  0  5    27,46,40                                     2, 9,27,13,46,40
 0 13 11    27,48,32,54,54, 8,26,15                      2, 9,12,16, 4,43,24, 6,54,48,53,20
29 22  0    27,51,46, 7,48, 9,15,43,29,16,48             2, 9, 3,31,27,45, 1,48,32,46,59,24,27,23,53,44,41,28,53,20
12 29  0    27,53,39,23,26,41,44,26,52,48                2, 8,59,40,13,54, 2,52,48
 9  0 16    27,54,29,23,34,48,53,20                      2, 8,53,29, 3,46,16,11, 7,40,24,34,21,54,50,37,30
51  0  3    27,55,49,45,36,42,58,57,46,40                2, 8,5C,56,28, 4,41, 5,16,48
 0 11 20    27,56,22,50,17, 0,37,47,34,41,15             2, 8,44,45,43, 3,59, 3,45
28  4  0    27,57,43,17,45,36                            2, 8,36, 2,57,46,40
11 11  0    27,59,36,57,36                               2, 8,27,20,47,51,55,11,45, 4,31,36,17,46,40
 0 21  3    28, 1,30,45, 8,26,15                         2, 8, 8,40,18,45
24  0  7    28, 5,35,48, 8,53,20                         2, 8
 1  4  4    28, 7,30                                     2, 7,51,20,16,27,39,15,33,20
 0 19 12    28, 9,24,19,35,19, 2,34,41,15                2, 7,27,55,31, 6,41,47,12,22,42,22,40,23,36, 2,39,29,16,22,42,57,46,40
10 34  0    28,14,34,37,59,16,45,45,12,57,3é             2, 7,24, 7, 8,32,38,24
 1  2 13    28,15,25,15,37,30                            2, 7,18, 0,33,21,15,14,41,39,10,11,43, 7,30
43  2  0    28,16,46,37,55,55,31,12                      2, 7, 5,23,40,18,45
26  9  0    28,18,41,35,13,55,12                         2, 7, C,47,22,14,58,45,55,33,20
 9 16  0    28,20,36,40,19,12                            2, 6,52,11,39, 7,34,30,51,55,34,55, 6,26,49,52,35,33,20
 0 27  4    28,22,31,53,12,17,34,41,15                   2, 6,48,24,20,34,37,21, 9,55, 7,23,31,12
 1  0 22    28,23,22,45, 6, 1,15,20,50                   2, 6,42,19,28, 8,55,41,36, 5,37,30
43  0  9    28,24,44,30,19,28,43,27,24,26,40             2, 6,33,45
12  0  2    28,26,40                                     2, 6,25,11, 6,40
 1 10  5    28,28,35,37,30                               2, 5,58,16,19,12
16  0 13    28,34,40,39,30,22,13,20                      2, 5,52,13,51, 1,35,29,37, 1,29,37,18,35,16,37,33,30,56,15
58  0  0    28,36, 2,57,15,59,51,15,29, 4                2, 5,49,44,49,55,12
 1  8 14    28,36,36,49,34,13, 7,30                      2, 5,43,42,46,16,32,50, 4, 6, 5,37,30
41  7  0    28,37,59,12,54,22,27,50,24                   2, 5,35,12,16, 6,40
24 14  0    28,39,55,36,25,20,38,24                      2, 5,26,42,20,29,36,33,30,25,30,51,51, 6,40
 7 21  0    28,41,52, 7,49,26,24                         2, 5,18,12,59,23, 2,14,11,17, 7, 4,47,50,56,40, 5,29,13, 5,11, 6,40
 0 33  5    28,43,48,47, 7,11,47,52,15,56,15             2, 5,14,28,29,12,42,49, 3, 7,46,33,36
 1  6 23    28,44,40,17, 9,50,46,17,20,37,30             2, 5, 8,28, 7, 3,37,58, 7,30
31  0  4    28,46, 3, 3,42,13,20                         2, 5
 6  3  0    28,48                                        2, 4,51,32,27,19,3C,22,13,20
 1 16  6    28,49,57, 4,13, 7,30                         2, 4,24,57,36
 4  0  8    28,56, 6,40                                  2, 4,18,59,36,19,20,58,52,51,50,44,15,23,43,49,41,15
56  5  0    28,57,29,59,28,56,51, 8,55,40,48             2, 4,1é,32,25,36
 1 14 15    28,58, 4,17,11,23,47,20,37,30
```

P	Q	R	
39	12	0	28,59,27,42,19, 3,14,41,16,48
22	19	0	29, 1,25,33, 7,39,38,52,48
5	26	0	29, 3,23,31,55,18,28,48
8	0	19	29, 4,15,37, 3,45,55,33,20
50	0	6	29, 5,39,20, 0,44,46,25,11, 6,4C
21	1	0	29, 7,37,36
4	8	0	29, 9,36
1	22	7	29,11,34,32, 1,17,20,37,30
23	0	10	29,15,49,47,39,15,33,20
0	4	7	29,17,48,45
37	17	0	29,21,12,18, 5,47,32, 7,17,45,36
20	24	0	29,23,11,37,17,30,23,36,57,36
3	31	0	29,25,11, 4,34,14,57,39,36
0	2	16	29,26, 3,48,46,33,45
38	0	1	29,27,28,34,30,45,20
19	6	0	29,29,28,19,12
2	13	0	29,31,28,12
1	28	8	29,33,28,12,55,18,18,37,58, 7,30
0	0	25	29,34,21,11,58,46,18,29,12, 5
11	0	5	29,37,46,40
0	10	8	29,39,47, 6,33,45
18	29	0	29,45,14, 1, 0,28,31,24,40,19,12
15	0	16	29,46, 7,21, 9, 8, 8,53,20
1	36	0	29,47,14,57,52,40,38,52,50,42
57	0	3	29,47,33, 4,39, 9,50,53,37,46,40
0	8	17	29,48, 8,21,38, 8,40,18,45
34	4	0	29,49,34,10,56,38,24
17	11	0	29,51,35,25,26,24
0	18	0	29,53,36,48, 9
30	0	7	29,57,58,11,21,28,53,20
1	1	1	30
0	16	9	30, 2, 1,56,53,40,18,45
3	0	11	30, 8,26,56,40
49	2	0	30, 9,53,44,27,39,13,16,48
0	14	18	30,10,29,27,54,22,16,48,59, 3,45
32	9	0	30,11,56,21,34,50,52,48
15	16	0	30,13,59, 7, 0,28,48
0	24	1	30,16, 2, 0,45, 6,45
7	0	22	30,16,56,16, 6,25,20,22,13,20
18	0	2	30,20,26,40
1	7	2	30,22,30
0	22	10	30,24,33,28,21,20,33,59, 3,45
22	0	13	30,28,59,22, 8,23,42,13,20
64	0	0	30,30,27, 9, 5, 3,50,40,31, 0,16
1	5	11	30,31, 3,16,52,30
47	7	0	30,32,31, 9,45,59,57,41,45,36
30	14	0	30,34,35,18,51, 2, 0,57,36
13	21	0	30,36,39,36,20,44, 9,36
0	30	2	30,38,44, 2,15,40,35, 3,45
1	3	20	30,39,38,58,18,30, 9,22,30
37	0	4	30,41, 7,15,57, 2,13,20
12	3	0	30,43,12
1	13	3	30,45,16,52,30
10	0	8	30,51,51, 6,40
1	11	12	30,53,56,34,20, 9,22,30
45	12	0	30,55,25,33, 8,19,27,40, 1,55,12
28	19	0	30,57,31,15,20,10,17,28,19,12
11	26	0	30,59,37, 6, 2,59,42,43,12
14	0	19	31, 0,32,39,32, 0,59,15,33,20
0	36	3	31, 1,43, 5,17,22,20,30, 2,48,45
1	9	21	31, 2,38,42,32,14, 1,59,31,52,30
27	1	0	31, 4, 8, 6,24
10	8	0	31, 6,14,24
1	19	4	31, 8,20,50, 9,22,30
29	0	10	31,12,53, 6,49,52,35,33,20
0	1	4	31,15
1	17	13	31,17, 7, 1,45,54,29,31,52,30
26	24	0	31,20,44,23,46,40,25,11,25,26,24
9	31	0	31,22,51,48,52,31,57,30,14,24
2	0	14	31,23,48, 4, 1,40
44	0	1	31,25,18,28,48,48,21,20
25	6	0	31,27,26,12,28,48
8	13	0	31,29,34, 4,48
1	25	5	31,31,42, 5,46,59,31,52,30
6	0	25	31,32,38,36,46,41,23,43, 8,53,2C
17	0	5	31,36,17,46,40
0	7	5	31,38,26,15
21	0	16	31,45,11,50,33,44,41,28,53,20
7	36	0	31,46,23,57,44,11,21,28,22, 4,48
0	5	14	31,47,20,55, 4,41,15
40	4	0	31,48,52,27,40,24,57,36
23	11	0	31,51, 1,47, 8, 9,36
6	18	0	31,53,11,15,21,36
1	31	6	31,55,20,52,21,19,46,31,24,22,30
0	3	23	31,56,18, 5,44,16,24,45,56,15
36	0	7	31,57,50, 4, 6,54,48,53,20
5	0	0	32
0	13	6	32, 2,10, 4,41,15
9	0	11	32, 9, 0,44,26,40
55	2	0	32,10,33,19,25,29,50, 9,55,12
0	11	15	32,11,11,25,45,59,45,56,15
38	9	0	32,12,44, 7, 1,10,16,19,12
21	16	0	32,14,55, 3,28,30,43,12
4	23	0	32,17, 6, 8,48, 7,12
13	0	22	32,18, 4, 1,10,51, 1,43,42,13,20
24	0	2	32,21,48,26,40
3	5	0	32,24
0	19	7	32,26,11,42,14,45,56,15
28	0	13	32,30,55,19,36,57,17, 2,13,20
1	2	8	32,33, 7,30

2, 4,1C,34,50, 8,56, 7,58, 7,30
2, 4, 2,10,38, 8, 3,57, 2,13,20
2, 3,53,47, 0,14,25,44,12,16,18,37,52,42, 8,23,42,13,20
2, 3,50, 5, 1,20,41, 9,53,16,48
2, 3,44, 8,42, 1,13, 8,16,57,59,35,23,26,15
2, 3,35,46,17,20,37,30
2, 3,27,24,26,40
2, 3,19, 3, 9,57, 2,35,16,52,20,44,26,40
2, 3, 1, 7,30
2, 2,52,48
2, 2,38,35,53,14, 0,37,30
2, 2,3C,17,54,42, 2,25,13,18,21,14, 4,76,40
2, 2,22, 0,29,52, 1,42,55, 4,59,52,57,58,39,24, 9, 6,30, 7,24,26,40
2, 2,18,21,15,24, 8, 3,50,24
2, 2,12,29,20, 1,12,14, 6,23,12,11,15
2, 2, 4,13, 7,30
2, 1,55,57,28,33,34,48,53,20
2, 1,47,42,23, 9,40,20, 1,50,57,31,18,11,21,28,53,20
2, 1,44, 4,1C, 9,14,15,31, 7,19, 5,46,45, 7,12
2, 1,3C
2, 1,21,46,4C
2, 0,59,33,14,45,57,56,45,44, 3,11,40,41, 9, 8, 8,53,20
2, 0,55,56,28, 1,55,12
2, 0,51,21,58,23,14,17,12,10,51,44, 9,51,15,57,11,13, 5,18,25,37,26,54,48,53,20
2, 0,5C, 8,29,47, 7,40,25,56,38, 2,13, 2,39,57,39,22,30
2, 0,47,45,26,19,23,31,12
2, 0,41,57,51,37,29, 7,15,56,15
2, 0,33,47,46,40
2, 0,25,38,14,52,25,29,46, 0,29,37,46,40
2, 0, 8, 7,47,34,41,15
2
1,59,51,52,45,25,55,33,2C
1,59,26,21,41,45,36
1,59,20,38, 1,16,1C,32,31,32,58,18,29,10,46,52,30
1,59,18,16,43,46,33,36
1,59,12,33,26,32,34,41,15
1,59, 4,29,24,36,32,35,33,20
1,58,56,25,55,25,51, 6,10,16,51,29, 9,47,39,15,33,20
1,58,52,52,49,17,27,31, 5,32,55,40,48
1,58,39, 8,26,15
1,58,31, 6,40
1,58,23, 5,26,21, 9,41, 4,11,51, 6,40
1,58, 5,52,48
1,58, C,12,59, 5,14,31,30,57,39, 1,13,40,34,20,12,40,15,14, 3,45
1,57,57,53,16,48
1,57,52,13,50,53, C,46,56,20,42,46,24,72,30
1,57,44,15,15, 6,15
1,57,36,17,11,42,45,31,24,46,25,11, 6,40
1,57,28,19,40,40,20,50,48, 4,47,53,14,51,30,37,35, 8,38,31, 6,40
1,57,24,49,12,23,10, 8,29,11, 2,24
1,57,19,11,21,37, 9,20,44,31,52,30
1,57,11,15
1,57, 3,19,10,37, 2,13,20
1,56,38,24
1,56,30,30,24
1,56,24,55, 9,30,52,37,28,14,31,52,30
1,56,17, 2,28,15, 3,42,13,20
1,56, 9,10,18,58,31,37,41,30,17,28, 0,79,30,22,13,20
1,56, 5,42,12,30,38,35,31,12
1,56, 1,18,41,39, 6,30,54,53,37,39,59,51,36,54,53,58, 9,53,41,23,57, 2,13,20
1,55,57,50,49,16,12,58,45, 7,12
1,55,52,17, 8,45,35, 9,22,30
1,55,44,26,40
1,55,36,36,43, 4,43,40,34,34, 4,26,40
1,55,19,48,16,52,30
1,55,12
1,55, 4,12,14,48,53,20
1,54,5C,54,17,31,54,46, 8,43,27,24,26,40
1,54,43, 7,58, 0, 1,36,29, 8,26, 8,24,24,71,14,26,23,32,20,44,26,40
1,54,39,42,25,41,22,33,36
1,54,34,12,30, 1, 7,43,13,29,15,10,32,48,45
1,54,26,27,18,16,52,30
1,54,18,42,38, 1,28,53,20
1,54,10,58,29,12,49, 3,46,44, 1,25,35,48, 8,53,20
1,54, 7,33,54,31, 9,37, 2,55,36,39,10, 4,48
1,54, 3,54,22,3C
1,53,46,40
1,53,22,26,41,16,48
1,53,18, 9,20,59,17, 8,37,40,11, 0, 9,14,18,42,21,46, 1,13,31,31,21,28,53,20
1,53,14,46,20,55,40,48
1,53, 9,20,29,38,53,33, 3,41,29, 3,45
1,53, 1,41, 2,30
1,52,54, 2, 6,26,38,54, 9,22,57,46,40
1,52,46,23,41,26,44, 0,46, 9,24,22,19, 3,51, 0, 4,56,17,46,40
1,52,43, 1,38,17,26,32, 8,48,59,54,14,24
1,52,37,37,18,21,16,10,18,45
1,52,3C
1,52,22,23,12,35,33,20
1,51,58,27,5C,24
1,51,53, 5,38,41,24,52,59,34,39,39,49,51,21,26,43, 7,30
1,51,50,53,11, 2,24
1,51,45,31,21, 8, 2,31,10,18,45
1,51,37,57,34,19,15,33,20
1,51,3C,24,18,12,59, 9,47, 2,40,46, 5,25,55,33,20
1,51,27, 4,31,12,37, 2,53,57, 7,12
1,51,14,11,39,36,33,45
1,51, 6,40
1,50,5S, 8,50,57,20,19,45,11, 6,40
1,50,43, 0,45
1,50,35,31,12

P	Q	R		
53	7	0	32,34,41,14,25, 3,57,32,32,38,24	1,50,30,12,58,57,11,59, 0,19,25, 6, 0,21, 5,37,30
0	17	16	32,35,19,49,20,19,15,45,42,11,15	1,50,28, 2, 9,25,20
36	14	0	32,36,53,40, 6,26, 9, 1,26,24	1,50,22,44,17,54,36,33,45
19	21	0	32,39, 6,14,46, 7, 6,14,24	1,50,15,16, 7,13,50,10,41,58,31, 6,40
2	28	0	32,41,18,58,24,43,17,24	1,50, 7,48,26,52,49,32,37,34,29,53,40,10,47,27,44,11,51, 6,40
1	0	17	32,42,17,34,11,44,10	1,50, 4,31, 7,51,43,15,27,21,36
43	0	4	32,43,51,45, 0,50,22,13,20	1,49,59,14,24, 1, 5, 0,41,44,52,58, 7,30
18	3	0	32,46, 4,48	1,49,51,47,48,45
1	10	0	32,48,18	1,49,44,21,43,42,13,20
0	25	8	32,50,31,21, 1,27, 0,42,11,15	1,49,36,56, 8,50,42,18, 1,39,51,46,10,22,13,20
16	0	8	32,55,18,31, 6,40	1,49,21
1	8	9	32,57,32,20,37,30	1,49,13,36
34	19	0	33, 1,21,20,21,30,58,38,12,28,48	1,49, 0,58,33,59, 7,13,20
17	26	0	33, 3,35,34,27,11,41,34, 4,48	1,48,53,35,55,17,22, 9, 5, 9,38,52,30,27, 2,13,20
20	0	19	33, 4,34,50,10, 9, 3,12,35,33,20	1,48,50,20,49,13,43,40,48
0	33	0	33, 5,49,57,38,31,49,52, 3	1,48,46,13,46,32,54,51,28,57,46,33,44,52, 8,21,28, 5,46,46,35, 3,42,13,20
1	6	18	33, 6,49,17,22,22,58, 7,30	1,48,42,58,53,41,27,10, 4,48
33	1	0	33, 8,24,38,49,36	1,48,37,46, 4,27,44,12,32,20,37,30
16	8	0	33,10,39,21,36	1,48,30,25
1	16	1	33,12,54,13,30	1,48,23, 4,25,23,10,56,47,24,26,40
35	0	10	33,17,44,39,17,12, 5,55,33,20	1,48, 7,19, 0,49,13, 7,30
4	0	3	33,20	1,48
1	14	10	33,22,15,29,52,58, 7,30	1,47,52,41,28,53,20
15	31	0	33,28,23,16, 8, 2, 5,20,15,21,35	1,47,32,56,13, 7,31,30,27,19, 9,30,22,49,54,47,14,34, 4,26,40
8	0	14	33,29,23,16,17,46,40	1,47,25,39,31,53,59,21,57,29,39,19,15,25,34,10,49,58,18, 3, 2,46,37,15,23,27,24,26,/
				/40
0	39	1	33,30,39,20, 6,45,43,44,27, 2,15	1,47,24,34,13, 8,33,29,16,23,40,28,38,15,42,11,15
50	0	1	33,30,59,42,44, 3,34,45,20	1,47,22,27, 3,23,54,14,24
1	12	19	33,31,39,24,20,24,45,21, 5,37,30	1,47,17,18, 5,53,19,13, 7,30
31	6	0	33,33,15,57,18,43,12	1,47,10, 2,28, 8,53,20
14	13	0	33,35,32,21, 7,12	1,47, 2,47,19,53,15,59,47,33,46,20,14,48,53,20
1	22	2	33,37,48,54,10, 7,30	1,46,47,13,35,37,30
23	0	5	33,42,42,57,46,40	1,46,40
0	4	2	33,45	1,46,32,46,53,43, 2,42,57,46,40
1	20	11	33,47,17,11,30,22,51, 5,37,30	1,46,17,17,31,12
27	0	16	33,52,12,37,55,59,40,14,48,53,20	1,46,10, 5,57, 7,12
0	2	11	33,54,30,18,45	1,46, 5, 0,27,47,42,42,14,42,38,29,45,56,15
46	4	0	33,56, 7,57,31, 6,37,26,24	1,45,57,49,43,35,37,30
29	11	0	33,58,25,54,16,42,14,24	1,45,50,39,28,32,28,58,16,17,46,40
12	18	0	34, 0,44, 0,23, 2,24	1,45,43,29,42,36,18,45,43,16,19, 5,55,22,21,33,49,37,46,40
1	28	3	34, 3, 2,15,50,45, 5,37,30	1,45,40,20,17, 8,51, 7,38,15,56, 9,36
0	0	20	34, 4, 3,18, 7,13,30,25	1,45,35,16,13,27,26,24,40, 4,41,15
42	0	7	34, 5,41,24,23,22,28, 8,53,20	1,45,28, 7,30
11	0	0	34, 8	1,45,20,59,15,33,20
0	10	3	34,10,18,45	1,44,58,33,36
15	0	11	34,17,36,47,24,26,40	1,44,53,31,32,31,19,34,40,51,14,41, 5,29,23,51,17,55,46,52,30
61	2	0	34,19,15,32,43,11,49,30,34,52,48	1,44,51,27,21,36
0	8	12	34,19,56,11,29, 3,45	1,44,46,25,38,33,47,21,43,25, 4,41,15
44	9	0	34,21,35, 3,29,14,57,24,28,48	1,44,39,20,13,25,33,20
27	16	0	34,23,54,43,42,24,46, 4,48	1,44,32,15,17, 4,40,27,55,21,15,43,12,35,33,20
10	23	0	34,26,14,33,23,19,40,48	1,44,25,10,49,29,11,51,49,24,15,53,59,52,27,13,24,34,20,54,19,15,33,20
1	34	4	34,28,34,32,32,38, 9,26,43, 7,30	1,44,22, 3,44,20,35,40,52,36,28,48
0	6	21	34,29,36,20,35,48,55,32,48,45	1,44,17, 3,25,53, 1,38,26,15
30	0	2	34,31,15,40,26,40	1,44,10
9	5	0	34,33,36	1,44, 2,57, 2,46,15,18,31, 6,40
0	16	4	34,35,56,29, 3,45	1,43,47,49,27,11,15
34	0	13	34,40,59, 0,55,25, 6,10,22,13,20	1,43,40,48
3	0	6	34,43,20	1,43,33,47, 1,20
0	14	13	34,45,41, 8,37,40,32,48,45	1,43,28,49, 1,47,26,46,38,26,15
42	14	0	34,47,21,14,46,51,53,37,32, 9,36	1,43,21,48,51,46,43,17,31,51, 6,40
25	21	0	34,49,42,39,45,11,34,39,21,36	1,43,14,49,10,12, 1,26,50,13,35,31,33,25, 6,59,45,11, 6,40
8	28	0	34,52, 4,14,18,22,10,33,36	1,43,11,44,11, 7,14,18,14,24
7	0	17	34,53, 6,44,28,31, 6,40	1,43, 6,47,15, 1, 0,56,54, 8,19,39,29,31,52,30
49	0	4	34,54,47,12, 0,53,43,42,13,20	1,42,59,48,34,27,11,15
24	3	0	34,57, 9, 7,12	1,42,52,50,22,13,20
7	10	0	34,59,31,12	1,42,45,52,38,17,32, 9,24, 3,37,17, 2,13,20
0	22	5	35, 1,53,26,25,32,48,45	1,42,30,56,15
22	0	8	35, 6,59,45,11, 6,40	1,42,24
1	5	6	35, 9,22,30	1,42,17, 4,13,10, 7,24,26,40
0	20	14	35,11,45,24,29, 8,48,13,21,33,45	1,42, 5,14,55,35, 2, 1, 1, 5,17,41,43,42,13,20
23	26	0	35,15,49,56,45, 0,28,20,21, 7,12	1,41,58,20,24,53,21,25,45,54, 9,54, 8,18,52,50, 7,35,25, 6,10,22,13,20
6	33	0	35,18,13,17,29, 5,57,11,31,12	1,41,55,17,42,50, 6,43,12
1	3	15	35,19,16,34,31,52,30	1,41,50,24,26,41, 0,11,45,19,20, 9,22,30
39	1	0	35,20,58,17,24,54,24	1,41,43,30,56,15
22	8	0	35,23,21,59, 2,24	1,41,36,37,53,47,59, 0,44,26,40
5	15	0	35,25,45,50,24	1,41,29,45,19,18, 3,36,41,32,27,56, 5, 9,27,54, 4,26,40
0	28	6	35,28, 9,51,30,21,58,21,33,45	1,41,26,43,28,27,41,52,55,56, 5,54,48,57,36
1	1	24	35,29,13,26,22,31,34,11, 2,30	1,41,21,51,34,31, 8,33,16,52,30
41	0	10	35,30,55,37,54,20,54,19,15,33,20	1,41,15
10	0	3	35,33,20	1,41, 8, 8,53,20
1	11	7	35,35,44,31,52,30	1,40,42,48,18,39,21,54,20, 9, 3, 6,48,12,43,17,39,20,54,25,21,21,12,25,40,44,26,40
14	0	14	35,43,20,49,22,57,46,40	1,40,41,47, 4,49,16,23,41,37,11,41,50,52,13,18, 2,48,45
4	38	0	35,44,41,57,27,12,46,39,24,50,24	1,40,39,47,51,56, 9,36
56	0	1	35,45, 3,41,34,59,49, 4,21,20	1,40,34,58,13, 1,14,16, 3,16,52,30
1	9	16	35,45,46, 1,57,46,24,22,30	1,40,28, 9,48,53,20
37	6	0	35,47,29, 1, 7,58, 4,48	1,40,21,21,52,23,41,14,48,20,24,41,28,53,20
20	13	0	35,49,54,30,31,40,48	1,40, 6,46,29,38,54,22,30
3	20	0	35,52,20, 9,46,48	1,40
29	0	5	35,57,33,49,37,46,40	1,39,53,13,57,51,36,17,46,40
2	2	0	36	1,39,31,58, 4,48
1	17	8	36, 2,26,20,16,24,22,30	1,39,27,11,41, 3,28,47, 6,17,28,35,24,18,59, 3,45
2	0	9	36,10, 8,20	1,39,25,13,56,28,48
52	4	0	36,11,52,29,21,11, 3,56, 9,36	1,39,20,27,52, 7, 8,54,22,30
1	15	17	36,12,35,21,29,14,44,10,46,52,30	1,39,13,44,30,30,27, 9,37,46,40
35	11	0	36,14,19,37,53,49, 3,21,36	1,39, 7, 1,36,11,32,35,21,49, 2,54,18, 9,42,42,57,46,40
18	18	0	36,16,46,56,24,34,33,36	1,39, 4, 4, 1, 4,32,55,54,37,26,24
1	25	0	36,19,14,24,54, 8, 6	1,38,59,18,57,36,58,30,37,34,23,40,18,45
6	0	20	36,20,19,31,19,42,24,26,40	1,38,52,37, 1,52,30
48	0	7	36,22, 4,10, 0,55,58, 1,28,53,20	1,38,45,55,33,20
17	0	0	36,24,32	
0	7	0	36,27	

P	Q	R	
1	23	9	36,29,28,10, 1,36,40,46,52,30
21	0	11	36,34,47,14,34, 4,26,40
0	5	9	36,37,15,56,15
50	9	0	36,39, 1,23,43,11,57,14, 6,43,12
33	16	0	36,41,30,22,37,14,25, 9, 7,12
16	23	0	36,43,59,31,36,52,59,31,12
1	31	1	36,46,28,50,42,48,42, 4,30
0	3	18	36,47,34,45,58,12,11,15
36	0	2	36,49,20,43, 8,26,40
15	5	0	36,51,50,24
0	13	1	36,54,20,15
0	1	27	36,57,56,29,58,27,53, 6,30, 6,15
9	0	6	37, 2,13,20
0	11	10	37, 4,43,53,12,11,15
31	21	0	37, 9, 1,30,24,12,20,57,59, 2,24
14	28	0	37,11,32,31,15,35,39,15,50,24
13	0	17	37,12,39,11,26,25,11, 6,40
1	37	2	37,14, 3,42,20,50,48,36, 3,22,30
55	0	4	37,14,26,20,48,57,18,37, 2,13,20
0	9	19	37,15,10,27, 2,40,50,23,26,15
30	3	0	37,16,57,43,40,48
13	10	0	37,19,29,16,48
0	19	2	37,22, 1, 0,11,15
28	0	8	37,27,27,44,11,51, 6,40
1	2	3	37,30
0	17	11	37,32,32,26, 7, 5,23,26,15
12	33	0	37,39,26,10,39, 2,21, 0,17,16,48
1	0	12	37,40,33,40,50
45	1	0	37,42,22,10,34,34, 1,36
28	8	0	37,44,55,26,58,33,36
11	15	0	37,47,28,53,45,36
0	25	3	37,50, 2,30,56,23,26,15
5	0	23	37,51,10,20, 8, 1,40,27,46,40
16	0	3	37,55,33,20
1	8	4	37,58, 7,30
0	23	12	38, 0,41,50,26,40,42,28,49,41,15
20	0	14	38, 6,14,12,40,29,37,46,40
62	0	1	38, 8, 3,56,21,19,48,20,38,45,20
1	6	13	38, 8,49, 6, 5,37,30
43	6	0	38,10,38,57,12,29,57, 7,12
26	13	0	38,13,14, 8,33,47,31,12
9	20	0	38,15,49,30,25,55,12
0	31	4	38,18,25, 2,49,35,43,49,41,15
1	4	22	38,19,33,42,53, 7,41,43, 7,30
35	0	5	38,21,24, 4,56,17,46,40
8	2	0	38,24
1	14	5	38,26,36, 5,37,30
8	0	9	38,34,48,53,20
58	4	0	38,36,39,59,18,35,48,11,54,14,24
1	12	14	38,37,25,42,55,11,43, 7,30
41	11	0	38,39,16,56,25,24,19,35, 2,24
24	18	0	38,41,54, 4,10,12,51,50,24
7	25	0	38,44,31,22,33,44,38,24
12	0	20	38,45,40,49,25, 1,14, 4,26,40
23	0	0	38,50,10, 8
6	7	0	38,52,48
1	20	6	38,55,26, 2,41,43, 7,30
27	0	11	39, 1, 6,23,32,20,44,26,40
0	2	6	39, 3,45
1	18	15	39, 6,23,47,12,23, 6,54,50,37,30
39	16	0	39, 8,16,24, 7,43,22,49,43,40,48
22	23	0	39,10,55,29,43,20,31,29,16,48
5	30	0	39,13,34,46, 5,39,56,52,48
0	0	15	39,16,45, 5, 2, 5
42	0	2	39,16,38, 6, 1, 0,26,40
21	5	0	39,19,17,45,36
4	12	0	39,21,57,36
1	26	7	39,24,37,37,13,44,24,50,37,30
4	0	26	39,25,48,15,58,21,44,38,56, 6,40
15	0	6	39,30,22,13,20
0	8	7	39,33, 2,48,45
20	28	0	39,40,18,41,20,38, 1,52,53,45,36
19	0	17	39,41,29,48,12,10,51,51, 6,40
3	35	0	39,42,59,57,10,14,11,50,27,36
0	6	16	39,44,11, 8,50,51,33,45
36	3	0	39,46, 5,34,35,31,12
19	10	0	39,48,47,13,55,12
2	17	0	39,51,29, 4,12
0	4	25	39,55,22,37,10,20,30,57,25,18,45
34	0	8	39,57,17,35, 8,38,31, 6,40
3	0	1	40
0	14	8	40, 2,42,35,51,33,45
7	0	12	40,11,15,55,33,20
1	40	0	40,12,47,12, 8, 6,52,29,20,26,42
51	1	0	40,13,11,39,16,52,17,42,24
0	12	17	40,13,59,17,12,29,42,25,18,45
34	8	0	40,15,55, 8,46,27,50,24
17	15	0	40,18,38,49,20,38,24
0	22	0	40,21,22,41, 0, 9
11	0	23	40,22,35, 1,28,33,47, 9,37,46,40
22	0	3	40,27,15,33,20
1	5	1	40,30
0	20	9	40,32,44,37,48,27,25,18,45
26	0	14	40,38,39, 9,31,11,36,17,46,40
1	3	10	40,41,24,22,30
49	6	0	40,43,21,33, 1,19,56,55,40,48
32	13	0	40,46, 7, 5, 8, 2,41,16,48
15	20	0	40,48,52,48,27,38,52,48
0	28	1	40,51,38,43, 0,54, 6,45
1	1	19	40,52,51,57,44,40,12,30

1,38,35,14,31,57,38, 4,13,29,52,35,33,20
1,38,24,54
1,38,14,24
1,38,13,31,32,24,10,39, 6,57,15,38,40,18,45
1,38, 6,52,42,35,12,30
1,38, C,14,19,45,37,56,10,38,40,59,15,33,20
1,37,53,36,23,53,37,22,20, 3,59,54,22,72,55,31,19,17,12, 5,55,33,20
1,37,5C,41, 0,19,18,27, 4,19,12
1,37,45,59,28, 0,57,47,17, 6,33,45
1,37,35,22,30
1,37,32,45,58,50,51,51, 6,40
1,37,23,15,20, 7,23,24,24,53,51,16,37,24, 5,45,36
1,37,12
1,37, 5,25,20
1,36,54,12, 3,32,33, 5,11, 6,40
1,36,47,38,35,48,46,21,24,35,14,33,20,32,55,18,31, 6,40
1,36,44,45,10,25,32, 9,36
1,36,41, 5,34,42,35,25,45,44,41,23,19,53, 0,45,44,58,28,14,44,29,57,31,51, 6,40
1,36,4C, 6,47,49,42, 8,20,45,18,25,46,26, 7,58, 7,30
1,36,38,12,21, 3,30,48,57,36
1,36,33,34,17,17,59,17,48,45
1,36,27, 2,13,20
1,36,20,30,35,53,56,23,48,48,23,42,13,20
1,36, 6,30,14, 3,45
1,36
1,35,53,30,12,20,44,26,40
1,35,35,56,38,20, 1,20,24,17, 1,47, 0,17,42, 1,59,36,57,17, 2,13,20
1,35,33, 5,21,24,28,48
1,35,28,30,25, 0,56,26, 1,14,22,38,47,20,37,30
1,35,22, 2,45,14, 3,45
1,35,15,35,31,41,14, 4,26,4C
1,35, 9, 8,44,20,40,53, 8,56,41,11,19,50, 7,24,26,40
1,35, 6,18,15,25,58, 0,52,26,20,32,38,24
1,34,55,18,45
1,34,48,53,20
1,34,42,28,21, 4,55,44,51,21,28,53,20
1,34,28,42,14,24
1,34,24,10,23,16,11,37,12,46, 7,12,58,56,27,28,10, 8,12,11,15
1,34,22,18,37,26,24
1,34,17,47, 4,42,24,37,33, 4,34,13, 7,70
1,34,11,24,12, 5
1,34, 5, 1,45,22,12,25, 7,49, 8, 8,53,70
1,33,5E,39,44,32,16,40,38,27,50,18,35,53,12,30, 4, 6,54,48,53,20
1,33,55,51,21,54,32, 6,47,20,49,55,12
1,33,51,21, 5,17,43,28,35,37,30
1,33,45
1,33,38,39,20,29,37,46,40
1,33,18,43,12
1,33,14,14,42,14,30,44, 9,38,53, 3,11,72,47,52,15,56,15
1,33,12,24,19,12
1,33, 7,56, 7,36,42, 5,58,35,37,30
1,33, 1,37,58,36, 2,57,46,40
1,32,55,20,15,10,49,18, 9,12,13,58,24,71,36,17,46,40
1,32,52,33,46, 0,30,52,24,57,36
1,32,41,49,43, 0,28, 7,30
1,32,35,33,20
1,32,29,17,22,27,46,56,27,39,15,33,20
1,32,15,50,37,30
1,32, 9,36
1,32, 3,21,47,51, 6,40
1,31,58,56,54,55,30,28, 7,30
1,31,52,43,26, 1,31,48,54,58,45,55,33,20
1,31,46,30,22,24, 1,17,11,18,44,54,43,78,59,33, 6,49,52,35,33,20
1,31,43,45,56,33, 6, 2,52,48
1,31,39,22, 0, 0,54,10,34,47,24, 8,26,15
1,31,33, 9,50,37,30
1,31,26,58, 6,25,11, 6,4C
1,31,20,46,47,22,15,15, 1,23,13, 8,28,38,31, 6,40
1,31,18, 3, 7,36,55,41,38,20,29,19,20, 3,50,24
1,31, 7,30
1,31, 1,20
1,30,44,39,56, 4,28,27,34,18, 2,23,45,70,51,51, 6,40
1,30,41,57,21, 1,26,24
1,30,38,31,28,47,25,42,54, 8, 8,48, 7,73,26,57,53,24,48,58,49,13, 5,11, 6,40
1,30,35,49, 4,44,32,38,24
1,30,31,28,23,43, 6,50,26,57,11,15
1,30,25,20,50
1,30,15,13,41, 9,19, 7,19,30,22,13,20
1,30,10,25,18,37,57,13,43, 3,11,55,23,71,12
1,30, 6, 5,50,41, 0,56,15
1,30
1,29,53,54,34, 4,26,40
1,29,34,46,16,19,12
1,29,31,22,56,34,59,28,17,54,42,46, 2,51,18,29, 1,38,35, 2,32,18,51, 2,49,32,50,22,/
1,29,30,28,30,57, 7,54,23,39,43,43,51,53, 5, 9,22,30   /13,20
1,29,28,42,32,49,55,12
1,29,24,25, 4,54,26, 0,56,15
1,29,18,22, 3,27,24,26,40
1,29,12,19,26,34,23,19,49,38, 8,36,52,20,44,26,40
1,29, 5,39,36,58, 5,38,19, 9,41,45,36
1,28,59,21,19,41,15
1,28,53,20
1,28,47,19, 4,45,52,15,48, 8,53,20
1,28,34,24,36
1,28,28,24,57,36
1,28,24,10,23, 9,45,35,12,15,32, 4,48,76,52,30
1,28,18,11,26,19,41,15
1,28,12,12,53,47, 4, 8,33,34,48,53,20
1,28, 6,14,45,30,15,38, 6, 3,35,54,56, 8,37,58,11,21,28,53,20
1,28, 3,36,54,17,22,36,21,53,16,48

P	Q	R		
41	0	5	40,54,49,41,16, 2,57,46,40	1,27,59,23,31,12,52, 0,33,23,54,22,30
14	2	0	40,57,36	1,27,53,26,15
1	11	2	41, 0,22,30	1,27,47,29,22,57,46,40
0	26	10	41, 3, 9,11,16,48,45,52,44, 3,45	1,27,41,32,55, 4,33,50,25,19,53,24,56,17,46,40
14	0	9	41, 9, 8, 8,53,20	1,27,28,48
1	9	11	41,11,55,25,46,52,30	1,27,22,52,48
47	11	0	41,13,54, 4,11, 5,56,53,22,23,36	1,27,18,41,22, 8, 9,28, 6,10,53,54,22,30
30	18	0	41,16,41,40,26,53,43,17,45,36	1,27,12,46,51,11,17,46,40
13	25	0	41,19,29,28, 3,59,36,57,36	1,27, 6,52,44,13,53,43,16, 7,43, 6, 0,?9,37,46,40
18	0	20	41,20,43,32,42,41,19, 0,44,26,4C	1,27, 4,16,39,22,58,56,38,24
0	34	2	41,22,17,27, 3, 9,47,20, 3,45	1,27, C,59, 1,14,19,53,11,10,13,14,59,53,42,41,10,28,37,25,16, 2,57,46,40
1	7	20	41,23,31,36,42,58,42,39,22,30	1,26,58,23, 6,57, 9,44, 3,50,24
29	0	0	41,25,30,48,32	1,26,54,12,51,34,11,22, 1,52,30
12	7	0	41,28,19,12	1,26,48,20
1	17	3	41,31, 7,46,52,30	1,26,42,27,32,18,32,45,25,55,33,20
33	0	11	41,37,10,49, 6,30, 7,24,26,40	1,26,29,51,12,39,22,30
2	0	4	41,40	1,26,24
1	15	12	41,42,49,22,21,12,39,22,30	1,26,18, 9,11, 6,40
28	23	0	41,47,39,11,42,13,53,35,13,55,12	1,26, 8,10,43, 8,56, 4,36,32,35,33,20
11	30	0	41,50,29, 5,10, 2,36,40,19,12	1,26, 2,20,58,30, 1,12,21,51,19,36,18,15,55,49,47,39,15,33,20
6	0	15	41,51,44, 5,22,13,20	1,25,59,46,49,16, 1,55,12
48	0	2	41,53,44,38,25, 4,28,26,40	1,25,55,39,22,30,5C,47,25, 6,56,22,54,36,33,45
27	5	0	41,56,34,56,38,24	1,25,49,50,28,42,39,22,30
10	12	0	41,59,25,26,24	1,25,44, 1,58,31, 6,40
1	23	4	42, 2,16, 7,42,39,22,30	1,25,38,13,51,54,36,47,5C, 3, 1, 4,11,51, 6,40
21	0	6	42, 8,23,42,13,20	1,25,25,46,52,30
0	5	4	42,11,15	1,25,20
1	21	13	42,14, 6,29,22,58,33,52, 1,52,3C	1,25,14,13,30,58,26,10,22,13,20
25	0	17	42,20,15,47,24,59,35,18,31, 6,40	1,25, 1,50, 0,57,36
9	35	0	42,21,51,56,58,55, 8,37,49,26,24	1,24,58,37, 0,44,27,51,28,15, 8,15, 6,55,44, 1,46,19,30,55, 8,38,31, 6,40
0	3	13	42,23, 7,53,26,15	1,24,56, 4,45,41,45,36
42	3	0	42,25, 9,56,53,53,16,48	1,24,52, 0,22,14,10, 9,47,46, 6,47,48,45
25	10	0	42,28, 2,22,50,52,48	1,24,46,15,46,52,30
8	17	0	42,30,55, 0,28,48	1,24,4C,31,34,49,59,10,37, 2,13,20
1	29	5	42,33,47,49,48,26,22, 1,52,30	1,24,34,47,46, 5, 3, 0,34,37, 3,16,44,17,53,15, 3,42,13,20
0	1	22	42,35, 4, 7,39, 1,53, 1,15	1,24,32,16,13,43, 4,54, 6,36,44,55,40,48
40	0	8	42,37, 6,45,29,13, 5,11, 6,40	1,24,28,12,58,45,57, 7,44, 3,45
9	0	1	42,40	1,24,22,30
0	11	5	42,42,53,26,15	1,24,16,47,24,26,4C
13	0	12	42,52, 0,59,15,33,20	1,23,58,50,52,48
57	1	0	42,54, 4,25,53,59,46,53,13,36	1,23,54,49,14, 1, 3,39,44,40,59,44,52,23,31, 5, 2,20,37,30
0	9	14	42,54,55,14,21,19,41,15	1,23,53, 9,53,16,48
40	8	0	42,56,58,49,21,33,41,45,36	1,23,49, 8,30,51, 1,53,22,44, 3,45
23	15	0	42,59,53,24,38, 0,57,36	1,23,43,28,10,44,26,40
6	22	0	43, 2,48,11,44, 9,36	1,23,37,48,13,39,44,22,2C,17, 0,34,34, 4,26,40
0	7	23	43, 7, 0,25,44,46, 9,26, 0,56,15	1,23,29,38,59,28,28,32,42, 5,11, 2,24
28	0	3	43, 9, 4,35,33,20	1,23,25,38,44,42,25,18,45
5	4	0	43,12	1,23,20
0	17	6	43,14,55,36,19,41,15	1,23,14,21,38,13, 0,14,48,53,20
32	0	14	43,21,13,46, 9,16,22,42,57,46,40	1,23, 2,15,53,45
1	0	7	43,24,10	1,22,56,38,24
55	6	0	43,26,14,59,13,25,16,43,23,31,12	1,22,52,39,44,12,53,59,15,14,33,49,30,15,49,13, 7,30
0	15	15	43,27, 6,25,47, 5,41, 0,56,15	1,22,51, 1,37, 4
38	13	0	43,29,11,33,28,34,52, 1,55,12	1,22,47, 3,23,57,25,18,45
21	20	0	43,32, 8,19,41,29,28,19,12	1,22,41,27, 5,25,22,38, 1,28,53,20
4	27	0	43,35, 5,17,52,57,43,12	1,22,35,51,20, 9,37, 9,28,10,52,25,15, 8, 5,35,48, 8,53,20
5	0	18	43,36,23,25,35,38,53,20	1,22,33,23,20,53,47,26,35,31,12
47	0	5	43,38,29, 0, 1, 7, 9,37,46,40	1,22,29,25,48, 0,48,45,31,18,39,43,35,37,30
20	2	0	43,41,26,24	1,22,23,50,51,33,45
3	9	0	43,44,24	1,22,18,16,17,46,40
0	23	7	43,47,21,48, 1,56, 0,56,15	1,22,12,42, 6,38, 1,43,31,14,53,49,37,46,40
20	0	9	43,53,44,41,28,53,20	1,22, C,45
1	6	8	43,56,43, 7,30	1,21,55,12
36	18	0	44, 1,48,27, 8,41,18,10,56,38,24	1,21,45,43,55,29,20,25
19	25	0	44, 4,47,25,56,15,35,25,26,24	1,21,40,11,56,28, 1,36,48,52,14, 9,22,57,46,40
2	32	0	44, 7,46,36,51,22,26,29,24	1,21,34,40,19,54,41, 8,36,43,19,55,18,?9, 6,16, 6, 4,20, 4,56,17,46,40
1	4	17	44, 9, 5,43, 9,50,37,30	1,21,32,14,10,56, 5,22,33,36
35	0	0	44,11,12,51,46, 8	1,21,28,19,33,20,48, 9,24,15,28, 7,30
18	7	0	44,14,12,28,48	1,21,22,48,45
1	14	0	44,17,12,18	1,21,17,18,19, 2,23,12,35,33,20
0	29	8	44,20,12,19,22,57,27,56,57,11,15	1,21,11,48,15,26,26,53,21,13,58,20,52, 7,34,19,15,33,20
1	2	26	44,21,31,47,58, 9,27,43,48, 7,3C	1,21, 5,22,46,46, 9,30,2C,44,52,43,51,10, 4,48
39	0	11	44,23,39,32,22,56, 7,54, 4,26,40	1,21, 5,29,15,36,54,50,37,30
8	0	4	44,26,40	1,21
1	12	9	44,29,40,39,50,37,30	1,20,54,31, 6,40
17	30	0	44,37,51, 1,30,42,47, 7, 0,28,48	1,20,35,42, 9,50,38,37,50,29,22, 7,47, 7,26, 5,25,55,33,20
12	0	15	44,39,11, 1,43,42,13,20	1,20,37,17,38,41,16,48
0	37	0	44,40,52,26,49, 0,58,19,16, 3	1,20,34,14,38,55,29,31,28, 7,14,29,26,34,10,38, 7,28,43,32,17, 4,57,56,32,35,33,20
54	0	2	44,41,19,36,58,44,46,20,26,40	1,20,33,25,39,51,25, 6,57,17,45,21,28,41,46,38,26,15
1	10	18	44,42,12,32,27,13, 0,28,·7,30	1,20,31,50,17,32,55,40,48
33	5	0	44,44,21,16,24,57,36	1,20,27,58,34,24,59,24,50,37,30
16	12	0	44,47,23, 8, 9,36	1,20,22,31,51, 6,40
1	20	1	44,50,25,12,13,30	1,20,17, 5,29,54,56,59,50,40,19,45,11, 6,40
27	0	6	44,56,57,17, 2,13,20	1,20, 5,25,11,43, 7,30
0	2	1	45	1,20
1	18	10	45, 3, 2,55,20,30,28, 7,30	1,19,54,35,10,17,17, 2,13,20
0	0	10	45,12,40,25	1,19,37,34,27,50,24
48	3	0	45,14,50,36,41,28,49,55,12	1,19,33,45,20,50,47, 1,41, 1,58,52,19,27,11,15
31	10	0	45,17,54,32,22,16,19,12	1,19,28,22,17,41,43, 7,30
14	17	0	45,20,58,40,30,43,12	1,19,22,59,36,24,21,43,42,13,20
1	26	2	45,24, 3, 1, 7,40, 7,30	1,19,17,37,16,57,14, 4,17,27,14,19,26,?1,46,10,22,13,20
4	0	21	45,25,24,24, 9,38, 0,33,20	1,19,15,15,12,51,38,20,43,41,57, 7,12
46	0	8	45,27,35,12,31, 9,57,31,51, 6,40	1,19,11,27,10, 5,34,48,30, 3,30,56,15
15	0	1	45,30,40	1,19, 6, 5,37,30
0	8	2	45,33,45	1,19, C,44,26,40
1	24	11	45,36,50,12,32, 0,50,58,35,37,3C	1,18,55,23,37,34, 6,27,22,47,54, 4,26,40
19	0	12	45,43,29, 3,12,35,33,20	1,18,43,55,12
63	1	0	45,45,40,43,37,35,46, 0,46,30,24	1,18,4C, 8,39,23,29,41, 0,38,26, 0,49, 7, 2,53,28,26,50.
0	6	11	45,46,34,55,18,45	1,18,38,35,31,12
46	8	0	45,48,46,44,38,59,56,32,38,24	1,18,34,49,13,55,20,31,17,33,48,30,56,15

P	Q	R	
29	15	0	45,51,52,58,16,33, 1,26,24
12	22	0	45,54,59,24,31, 6,14,24
1	32	3	45,58, 6, 3,23,30,52,35,37,30
0	4	20	45,59,28,27,27,45,14, 3,45
34	0	3	46, 1,40,53,55,33,20
11	4	0	46, 4,48
0	14	3	46, 7,55,18,45
7	0	7	46,17,46,40
0	12	12	46,20,54,51,30,14, 3,45
44	13	0	46,23, 8,19,42,29,11,30, 2,52,48
27	20	0	46,26,16,53, 0,15,26,12,28,48
10	27	0	46,29,25,39, 4,29,34, 4,48
11	0	18	46,30,48,59,18, 1,28,53,20
53	0	5	46,33, 2,56, 1,11,38,16,17,46,40
0	10	21	46,33,58, 3,48,21, 2,59,17,48,45
26	2	0	46,36,12, 9,36
9	9	0	46,39,21,36
0	20	4	46,42,31,15,14, 3,45
26	0	9	46,49,19,40,14,48,53,20
1	3	5	46,52,30
0	18	13	46,55,40,32,38,51,44,17,48,45
25	25	0	47, 1, 6,35,40, 0,37,47, 8, 9,36
8	32	0	47, 4,17,43,18,47,56,15,21,36
1	1	14	47, 5,42, 6, 2,30
41	0	0	47, 7,57,43,13,12,32
24	7	0	47,11, 9,18,43,12
7	14	0	47,14,21, 7,12
0	26	5	47,17,33, 8,40,29,17,48,45
3	0	24	47,18,57,55,10, 2, 5,34,43,20
14	0	4	47,24,26,40
1	9	6	47,27,39,22,30
18	0	15	47,37,47,45,50,37, 2,13,20
6	37	0	47,39,35,56,36,17, 2,12,33, 7,12
60	0	2	47,40, 4,55,26,39,45,25,48,26,4C
1	7	15	47,41, 1,22,37, 1,52,30
39	5	0	47,43,18,41,30,37,26,24
22	12	0	47,46,32,40,42,14,24
5	19	0	47,49,46,53, 2,24
0	32	6	47,53, 1,18,31,59,39,47, 6,33,45
1	5	24	47,54,27, 8,36,24,37, 8,54,22,30
33	0	6	47,56,45, 6,10,22,13,20
4	1	0	48
1	15	7	48, 3,15, 7, 1,52,30
6	0	10	48,13,31, 6,40
54	3	0	48,15,49,59, 8,14,45,14,52,48
1	13	16	48,16,47, 8,38,59,38,54,22,30
37	10	0	48,19, 6,10,31,45,24,28,48
20	17	0	48,22,22,35,12,46, 4,48
3	24	0	48,25,39,13,12,10,48
10	0	21	48,27, 6, 1,46,16,32,35,33,20
21	0	1	48,32,42,40
2	6	0	48,36
1	21	8	48,39,17,33,22, 8,54,22,30
25	0	12	48,46,22,59,25,25,55,33,20
0	3	8	48,49,41,15
52	8	0	48,52, 1,51,37,35,56,18,48,57,36
35	15	0	48,55,20,30, 9,39,13,32, 9,36
18	22	0	48,58,39,22, 9,10,39,21,36
1	29	0	49, 1,58,27,37, 4,56, 6
0	1	17	49, 3,26,21,17,36,15
40	0	3	49, 5,47,37,31,15,33,20
17	4	0	49, 9, 7,12
0	11	0	49,12,27
1	27	9	49,15,47, 1,32,10,31, 3,16,52,30
2	0	27	49,17,15,19,57,57,10,48,40, 8,20
13	0	7	49,22,57,46,40
0	9	9	49,26,18,30,56,15
33	20	0	49,32, 2, 0,32,16,27,57,18,43,12
16	27	0	49,35,23,21,40,47,32,21, 7,12
17	0	18	49,36,52,15,15,13,34,48,53,20
1	35	1	49,38,44,56,27,47,44,48, 4,30
0	7	18	49,40,13,56, 3,34,27,11,15
32	2	0	49,42,36,58,14,24
15	9	0	49,45,59, 2,24
0	17	1	49,49,21,20,15
32	0	9	49,56,36,58,55,48, 8,53,20
1	0	2	50
0	15	10	50, 3,23,14,49,27,11,15
14	32	0	50,12,34,54,12, 3, 8, 0,23, 2,24
5	0	13	50,14, 4,54,26,40
47	0	0	50,16,29,34, 6, 5,22, 8
0	13	19	50,17,29, 6,30,37, 8, 1,38,26,15
30	7	0	50,19,53,55,58, 4,48
13	14	0	50,23,18,31,40,48
0	23	2	50,26,43,21,15,11,15
9	0	24	50,28,13,46,50,42,13,57, 2,13,20
20	0	4	50,34, 4,26,40
1	6	3	50,37,30
0	21	11	50,40,55,47,15,34,16,38,26,15
24	0	15	50,48,18,56,53,59,30,22,13,20
1	4	12	50,51,45,28, 7,30
45	5	0	50,54,11,56,16,39,56, 9,3€
28	12	0	50,57,38,51,25, 3,21,36
11	19	0	51, 1, 6, 0,34,33,36
0	29	3	51, 4,33,23,46, 7,38,26,15
1	2	21	51, 6, 4,57,10,50,15,37,30
39	0	6	51, 8,32, 6,35, 3,42,13,20
10	1	0	51,12
1	12	4	51,15,28, 7,30
12	0	10	51,26,25,11, 6,40

1,18,24,30,10, 4,10
1,18,24,11,27,48,30,20,56,30,56,47,24,76,40
1,18,18,53, 7, 6,53,53,52, 3,11,55,29,54,20,25, 3,25,45,40,44,26,40
1,18,16,32,48,15,26,45,39,27,21,36
1,18,12,47,34,24,46,13,49,41,15
1,18, 7,30
1,18, 2,12,47, 4,41,28,53,20
1,17,45,36
1,17,40,20,16
1,17,36,36,46,20,35, 4,58,49,41,15
1,17,31,21,38,50, 2,28, 8,53,20
1,17,26, 6,52,39, 1, 5, 7,40,11,38,40,76,20,14,48,53,20
1,17,23,48, 8,20,25,43,40,48
1,17,20, 5,26,15,45,42,40,36,14,44,37, 8,54,22,30
1,17,18,33,52,50,48,39,10, 4,48
1,17,14,51,25,50,23,26,15
1,17, 9,37,46,40
1,17, 4,24,28,43, 9, 7, 3, 2,42,57,46,40
1,16,53,12,11,15
1,16,48
1,16,42,48, 9,52,35,33,20
1,16,33,56,11,41,16,30,45,48,58,16,17,46,40
1,16,26,45,18,40, 1, 4,19,25,37,25,36,14, 9,37,35,41,33,49,37,46,40
1,16,26,28,17, 7,35, 2,24
1,16,22,48,20, 0,45, 8,48,59,30, 7, 1,52,30
1,16,17,38,12,11,15
1,16,12,28,25,20,59,15,33,20
1,16, 7,18,59,28,32,42,31, 9,20,57, 3,52, 5,55,33,20
1,16, 5, 2,36,20,46,24,41,57, 4,26, 6,43,12
1,15,56,15
1,15,51, 6,40
1,15,34,57,47,31,12
1,15,32, 6,13,59,31,25,45, 6,47,20, 6, 9,32,28,14,30,40,49, 1, 0,54,19,15,33,20
1,15,31,20,18,36,57,17,46,12,53,46,23, 9, 9,58,32, 6,33,45
1,15,26,50,53,57, 7,12
1,15,26,13,39,45,55,42, 2,27,39,22,30
1,15,21, 7,21,40
1,15,16, 1,24,17,45,56, 6,15,18,31, 6,40
1,15,10,55,47,37,49,20,30,46,16,14,52,42,34, 0, 3,17,31,51, 6,40
1,15, 8,41, 5,31,37,41,25,52,39,56, 9,36
1,15, 5, 4,52,14,10,46,52,30
1,15
1,14,54,55,28,23,42,13,20
1,14,3€,58,33,36
1,14,35,23,45,47,36,35,19,43, 6,26,33,14,14,17,48,45
1,14,33,55,27,21,36
1,14,30,20,54, 5,21,40,40,46,52,30
1,14,25,18,22,52,50,22,13,20
1,14,20,16,12, 8,39,26,31,21,47,10,43,?7,17, 2,13,20
1,14,18, 3, 0,48,24,41,55,58, 4,48
1,14, 5,27,46,24,22,30
1,14, 4,26,40
1,13,59,25,53,58,13,33,10, 7,24,26,40
1,13,48,40,30
1,13,43,40,48
1,13,4C, 8,39,18, 7,59,20,12,56,44, 0,14, 3,45
1,13,35, 9,31,56,24,22,30
1,13,3C,10,44,49,13,27, 7,59, 0,44,26,40
1,13,25,12,17,55,13, 1,45, 2,59,55,46,47,11,38,29,27,54, 4,26,40
1,13,23, 0,45,14,28,50,18,14,24
1,13,19,29,36, 0,43,20,27,49,55,18,45
1,13,14,31,52,30
1,13, 9,34,29, 8, 8,53,20
1,13, 4,37,25,53,48,12, 1, 6,34,30,46,54,48,53,20
1,13, 2,26,30, 5,32,33,18,40,23,27,28, 3, 4,19,12
1,12,54
1,12,49, 4
1,12,40,39, 2,39,24,48,53,20
1,12,35,43,56,51,34,46, 3,26,25,55, 0,24,41,28,53,20
1,12,33,33,52,49, 9, 7,12
1,12,30,49,11, 1,56,34,19,18,31, 2,29,54,45,34,18,43,51,11, 3,22,28, 8,53,20
1,12,28,39,15,47,38, 6,43,12
1,12,25,10,42,58,29,28,21,33,45
1,12,20,16,40
1,12,15,22,56,55,27,17,51,36,17,46,40
1,12, 4,52,40,32,48,45
1,12
1,11,55, 7,39,15,33,20
1,11,41,57,28,45, 1, 0,18,12,46,20,15,13,16,31,29,42,42,57,46,40
1,11,39,49, 1, 3,21,36
1,11,36,22,48,45,42,19,30,55,46,59, 5,30,28, 7,30
1,11,34,58, 2,15,56, 9,36
1,11,31,32, 3,55,32,48,45
1,11,26,41,38,45,55,33,20
1,11,21,51,33,15,30,39,51,42,30,53,29,52,35,33,20
1,11,19,43,41,34,28,30,39,19,45,24,28,48
1,11,11,29, 3,45
1,11, 6,40
1,11, 1,51,15,48,41,48,38,31, 6,40
1,10,51,31,40,48
1,10,46,43,58, 4,48
1,10,43,20,18,31,48,28, 9,48,25,39,50,37,30
1,10,38,33, 9, 3,45
1,10,33,46,19, 1,39,18,50,51,51, 6,40
1,10,28,59,48,24,12,30,28,52,43,56,54,54,22,33, 5,11, 6,40
1,10,26,53,31,25,54, 5, 5,30,37,26,24
1,10,23,30,48,58,17,36,26,43, 7,30
1,10,18,45
1,10,13,59,30,22,13,20
1, 9,59, 2,24

P	Q	R		
60	3	0	51,28,53,19, 4,47,44,15,52,19,12	1, 9,55,41, 1,40,53, 3, 7,14, 9,47,23,39,35,54,11,57,11,15
1	10	13	51,29,54,17,13,35,37,30	1, 9,54,18,14,24
43	10	0	51,32,22,35,13,52,26, 6,43,12	1, 9,50,57, 5,42,31,34,28,56,43, 7,30
26	17	0	51,35,52, 5,33,37, 9, 7,12	1, 9,46,13,28,57, 2,13,20
9	24	0	51,39,21,50, 4,59,31,12	1, 9,41,30,11,23, 6,58,36,54,10,28,48,23,42,13,20
16	0	21	51,40,54,25,53,21,38,45,55,23,20	1, 9,39,25,19,30,23, 9,18,43,12
0	35	4	51,42,51,48,48,57,14,10, 4,41,15	1, 9,36,47,12,59,27,54,32,56,10,35,59,54,58, 8,56,22,53,56,12,50,22,13,20
1	8	22	51,44,24,30,53,43,23,19,13, 7,30	1, 9,34,42,29,33,43,47,15, 4,19,12
27	0	1	51,46,53,30,40	1, 9,31,22,17,15,21, 5,37,30
8	6	0	51,50,24	1, 9,26,40
1	18	5	51,53,54,43,35,37,30	1, 9,21,58, 1,50,50,12,20,44,26,40
31	0	12	52, 1,28,31,23, 7,39,15,33,20	1, 9,11,52,58, 7,30
0	0	5	52, 5	1, 9, 7,12
1	16	14	52, 8,31,42,56,30,49,13, 7,30	1, 8,59,12,41,11,37,51, 5,37,30
41	15	0	52,11, 1,52,10,17,50,26,18,14,24	1, 8,54,32,34,31, 8,51,41,14, 4,26,40
24	22	0	52,14,33,59,37,47,21,59, 2,24	1, 8,49,52,46,48, 0,57,53,29, 3,41, 2,76,44,39,50, 7,24,26,40
7	29	0	52,18, 6,21,27,33,15,50,24	1, 8,47,49,27,24,49,32, 9,36
4	0	16	52,19,40, 6,42,46,40	1, 8,44,31,30, 0,40,37,56, 5,33, 6,19,41,15
46	0	3	52,22,10,48, 1,20,35,33,20	1, 8,35,52,22,58, 7,30
23	4	0	52,25,43,40,48	1, 8,35,13,34,48,53,20
6	11	0	52,29,16,48	1, 8,30,35, 5,31,41,26,16, 2,24,51,21,28,53,20
1	24	6	52,32,50, 9,38,19,13, 7,30	1, 8,20,37,30
19	0	7	52,40,29,37,46,40	1, 8,16
0	6	6	52,44, 3,45	1, 8, 3,29,57, 3,21,20,40,43,31,47,49, 8, 8,53,20
22	27	0	52,53,44,55, 7,30,42,30,31,40,48	1, 8, 1,28, 0,46, 4,48
23	0	18	52,55,19,44,16,14,29, 8, 8,53,20	1, 7,58,53,36,35,34,17,10,36, 6,36, 5,92,35,13,25, 3,36,44, 6,54,48,53,20
5	34	0	52,57,19,56,13,38,55,47,16,48	1, 7,56,51,48,33,24,28,48
0	4	15	52,58,54,51,47,48,45	1, 7,53,36,17,47,20, 7,50,12,53,26,15
38	2	0	53, 1,27,26, 7,21,36	1, 7,45, 0,37,30
21	9	0	53, 5, 2,58,33,36	1, 7,44,25,15,51,59,20,29,37,46,40
4	16	0	53, 8,38,45,36	1, 7,39,50,12,52, 2,24,27,41,38,37,23,26,18,36, 2,57,46,40
1	30	7	53,12,14,47,15,32,57,32,20,37,30	1, 7,37,48,58,28,27,55,17,17,23,56,32,38,24
0	2	24	53,13,50, 9,33,47,21,16,33,45	1, 7,34,34,23, 0,45,42,11,15
38	0	9	53,16,23,26,51,31,21,28,53,20	1, 7,30
7	0	2	53,20	1, 7,25,25,55,33,20
0	12	7	53,23,36,47,48,45	1, 7,11, 4,42,14,24
11	0	13	53,35, 1,14, 4,26,40	1, 7, 8,32,12,26,14,36,13,26, 2, 4,32, 8,28,51,46,13,56,16,54,14, 8,17, 7, 9,37,
3	39	0	53,37, 2,56,10,49, 9,59, 7,15,36	1, 7, 7,51,23,12,50,55,47,44,47,47,53,54,48,52, 1,52,30 /40
53	0	0	53,37,35,32,22,29,43,36,32	1, 7, 6,31,54,37,26,24
0	10	16	53,38,39, 2,56,39,36,33,45	1, 7, 3,18,48,40,49,30,42,11,15
36	7	0	53,41,13,31,41,57, 7,12	1, 6,58,46,32,35,33,20
19	14	0	53,44,51,45,47,31,12	1, 6,54,14,34,55,47,29,52,13,36,27,39,15,33,20
2	21	0	53,48,30,14,40,12	1, 6,44,30,59,45,56,15
26	0	4	53,56,20,44,26,40	1, 6,40
1	3	0	54	1, 6,35,29,18,34,24,11,51, 6,40
0	18	8	54, 3,39,30,24,36,33,45	1, 6,25,48,27
30	0	15	54,11,32,12,41,35,28,23,42,13,20	1, 6,21,18,43,12
1	1	9	54,15,12,30	1, 6,18, 7,47,22,19,11,24,11,39, 3,36,12,39,22,30
51	5	0	54,17,48,44, 1,46,35,54,14,24	1, 6,16,49,17,39,12
0	16	17	54,18,53, 2,13,52, 6,16,10,18,45	1, 6,13,38,34,44,45,56,15
34	12	0	54,21,29,26,50,43,35, 2,24	1, 6, 9, 9,40,20,18, 6,25,11, 6,40
17	19	0	54,25,10,24,36,51,50,24	1, 6, 4,41, 4, 7,41,43,34,32,41,56,12, 6,28,28,38,31, 6,40
0	26	0	54,28,51,37,21,12, 9	1, 6, 2,42,40,43, 1,57,16,24,57,36
3	0	19	54,30,29,16,59,33,36,40	1, 5,59,32,38,24,39, 0,25, 2,55,46,52,30
45	0	6	54,33, 6,15, 1,23,57, 2,13,20	1, 5,55, 4,41,15
16	1	0	54,36,48	1, 5,50,37, 2,13,20
1	9	1	54,40,30	1, 5,46, 9,41,18,25,22,48,59,55, 3,42,13,20
0	24	9	54,44,12,15, 2,25, 1,10,18,45	1, 5,36,36
18	0	10	54,52,10,51,51, 6,40	1, 5,32, 9,36
1	7	10	54,55,53,54,22,30	1, 5,29, 1, 1,36, 7, 6, 4,38,10,25,46,52,30
49	10	0	54,58,32, 5,34,47,55,51,10, 4,48	1, 5,24,35, 8,23,28,20
32	17	0	55, 2,15,33,55,51,37,43,40,48	1, 5,20, 9,33,10,25,17,27, 5,47,19,30,22,13,20
15	24	0	55, 5,59,17,25,19,29,16,48	1, 5,15,44,15,55,44,56,53,22,39,56,14,55,17, 0,52,51,28, 3,57, 2,13,20
0	32	2	55, 9,43,16, 4,13, 3, 6,45	1, 5,13,47,20,12,52,18, 2,52,48
1	5	19	55,11,22, 8,57,18,16,52,30	1, 5,1C,39,38,40,38,31,31,24,22,30
33	0	1	55,14, 1, 4,42,40	1, 5, 6,15
14	6	0	55,17,45,36	1, 5, 1,50,39,13,54,34, 4,26,40
1	15	2	55,21,30,22,30	1, 4,52,23,24,29,31,52,30
37	0	12	55,29,34,25,28,40, 9,52,35,33,20	1, 4,48
6	0	5	55,33,20	1, 4,43,36,53,20
1	13	11	55,37, 5,49,48,16,52,30	1, 4,36, 8, 2,21,42, 3,27,24,26,40
30	22	0	55,43,32,15,36,18,31,26,58,23,36	1, 4,31,45,43,52,30,54,16,23,29,42,13,41,56,52,20,44,26,40
13	29	0	55,47,18,46,53,23,28,53,45,36	1, 4,29,50, 6,57, 1,26,24
10	0	16	55,48,58,47, 9,37,46,40	1, 4,27,23,43, 8,23,37,10,29,47,35,33,15,20,30,29,58,58,49,49,39,58,21,14, 4,26,40
0	38	2	55,51, 5,33,31,16,12,54, 5, 3,45	1, 4,26,44,31,53, 8, 5,33,50,12,17,10,57,25,18,45
52	0	3	55,51,39,31,13,25,57,55,33,20	1, 4,25,28,14, 2,20,32,38,24
1	11	20	55,52,45,40,34, 1,15,35, 9,22,30	1, 4,22,22,51,31,59,31,52,30
29	4	0	55,55,26,35,31,12	1, 4,18, 1,28,53,20
12	11	0	55,59,13,55,12	1, 4,13,40,23,55,57,35,52,32,15,48, 8,53,20
1	21	3	56, 3, 1,30,16,52,30	1, 4, 4,20, 9,22,30
25	0	7	56,11,11,36,17,46,40	1, 4
0	3	3	56,15	1, 3,55,40, 8,13,49,37,46,40
1	19	12	56,18,48,39,10,38, 5, 9,22,30	1, 3,43,57,45,33,20,53,36,11,21,11,20,11,48, 1,19,44,38,11,21,28,53,20
11	34	0	56,29, 9,15,58,33,31,30,25,55,12	1, 3,42, 3,34,16,19,12
0	1	12	56,30,50,31,15	1, 3,39, 0,16,4C,37,37,20,49,35, 5,51,33,45
44	2	0	56,33,33,15,51,51, 2,24	1, 3,34,41,50, 7,29,22,57,46,40
27	9	0	56,37,23,10,27,50,24	1, 3,30,23,41, 7,29,22,57,46,40
10	16	0	56,41,13,20,38,24	1, 3,26, 5,49,33,47,15,25,57,47,27,33,13,24,56,17,46,40
1	27	4	56,45, 3,46,24,35, 9,22,30	1, 3,24,12,10,17,18,40,34,57,33,41,45,36
2	0	22	56,46,45,30,12, 2,30,41,40	1, 3,21, 9,44, 4,27,50,48, 2,48,45
44	0	9	56,49,29, 0,38,57,26,54,48,53,20	1, 3,16,52,30
13	0	2	56,53,20	1, 3,12,35,33,20
0	4	4	56,57,11,15	1, 2,59, 8, 9,36
17	0	13	57, 9,21,19, 0,44,26,40	1, 2,56, 6,55,30,47,44,48,30,44,48,39,17,38,18,46,45,28, 7,30
59	0	0	57,12, 5,54,31,59,42,30,58, 8	1, 2,54,52,24,57,36
0	7	13	57,13,13,39, 8,26,15	1, 2,51,51,23, 8,16,25, 2, 3, 2,48,45
42	7	0	57,15,58,25,48,44,55,40,48	1, 2,47,36, 8, 3,20
25	14	0	57,19,51,12,50,41,16,48	1, 2,43,21,10,14,48,16,45,12,45,25,55,33,20
8	21	0	57,23,44,15,38,52,48	1, 2,39, 6,29,41,31, 7, 5,38,33,32,23,55,28,20, 2,44,36,32,35,33,20
1	33	5	57,27,37,34,14,23,35,44,31,52,30	

```
 P   Q   R
 0   5  22   57,29,20,34,,19,41,32,34,41,15        1, 2,37,14,14,36,21,24,31,33,53,16,48
32   0   4   57,32, 6, 7,24,26,40                  1, 2,34,14, 3,31,48,59, 3,45
 7   3   0   57,36                                 1, 2,30
 0  15   5   57,39,54, 8,26,15                      1, 2,25,46,13,39,45,11, 6,40
 5   0   8   57,52,13,20                           1, 2,12,28,48
57   5   0   57,54,59,58,57,53,42,17,51,21,36       1, 2, 5,29,48, 9,40,29,26,25,55,22, 7,41,51,54,50,37,30
 0  13  14   57,56, 8,34,22,47,34,41,15             1, 2, 8,16,12,48
40  12   0   57,58,55,24,38, 6,29,22,33,36          1, 2, 5,17,25, 4,28, 3,59, 3,45
23  19   0   58, 2,51, 6,15,19,17,45,36             1, 2, 1, 5,19, 4, 1,58,31, 6,40
 6  26   0   58, 6,47, 3,50,36,57,36                1, 1,56,53,30, 7,12,52, 6, 8, 9,18,56,21, 4,11,51, 6,40
 9   0  19   58, 8,31,14, 7,31,51, 6,40             1, 1,55, 2,30,40,2C,34,56,38,24
51   0   6   58,11,18,40, 1,29,32,50,22,13,20       1, 1,52, 4,21, 0,36,34, 8,28,59,47,41,43, 7,30
22   1   0   58,15,15,12                           1, 1,47,53, 8,40,18,45
 5   8   0   58,19,12                              1, 1,43,42,13,2C
 0  21   6   58,23, 9, 4, 2,34,41,15                1, 1,39,31,34,58,31,17,38,26,10,22,13,20
24   0  10   58,31,39,35,18,31, 6,40                1, 1,3C,33,45
 1   4   7   58,35,37,30                           1, 1,26,24
 0  19  15   58,39,35,40,48,34,40,22,15,56,15       1, 1,22,14,31,54, 4,26,4C
38  17   0   58,42,24,36,11,35, 4,14,35,31,12       1, 1,19,17,56,37, 0,18,45
21  24   0   58,46,23,14,35, 0,47,13,55,12          1, 1,15, 8,57,21, 1,12,36,39,10,37, 2,'3,20
 4  31   0   58,50,22, 9, 8,29,55,19,12             1, 1,11, 0,14,56, 0,51,27,32,29,56,28,59,19,42, 4,33,15, 3,42,13,20
 1   2  16   58,52, 7,37,33, 7,30                   1, 1, 9,10,37,42, 4, 1,55,12
39   0   1   58,54,57, 9, 1,30,40                   1, 1, 6,14,40, 0,36, 7, 3,11,36, 5,37,30
20   6   0   58,58,56,38,24                        1, 1, 2, 6,33,45
 3  13   0   59, 2,56,24                           1, 0,57,58,44,16,47,24,26,4C
 0  27   7   59, 6,56,25,50,36,37,15,56,15          1, 0,53,51,11,34,50,10, 0,55,28,45,39, 5,40,44,26,40
 1   0  25   59, 8,42,23,57,32,36,58,24,10          1, 0,52, 2, 5, 4,37, 7,45,33,39,32,53,22,33,36
12   0   5   59,15,33,20                           1, 0,45
 1  10   8   59,19,34,13, 7,30                      1, 0,4C,53,20
19  29   0   59,30,28, 2, 0,57, 2,49,20,38,24       1, 0,25,46,37,22,58,58,22,52, 1,35,50,20,34,34, 4,26,40
16   0  16   59,32,14,42,18,16,17,46,40             1, 0,27,58,14, C,57,36
 2  36   0   59,34,29,55,45,21,17,45,41,24          1, 0,25,40,59,11,37, 8,36, 5,25,52, 4,55,37,58,35,36,32,39,12,48,43,27,24,26,40
58   0   3   59,35, 6, 9,18,19,41,47,15,33,20       1, 0,25, 4,14,53,33,50,12,58,19, 1, 6,'1,19,58,49,41,15
 1   8  17   59,36,16,43,16,17,20,37,30             1, 0,23,52,43, 9,41,45,36
35   4   0   59,39, 8,21,53,16,48                   1, 0,20,58,55,48,44,33,37,58, 7,30
18  11   0   59,43,10,50,52,48                      1, 0,16,53,53,20
 1  18   0   59,47,13,36,18                         1, 0,12,49, 7,26,12,44,53, 0,14,48,53,70
31   0   7   59,55,56,22,42,57,46,40                1, 0, 4, 3,53,47,2C,37,30
 0   0   0   60                                    1
```

www.ingramcontent.com/pod-product-compliance
Lightning Source LLC
Chambersburg PA
CBHW081332190326

41458CB00018B/5975

*9 7 8 1 4 2 2 3 7 6 1 1 9 *